AGRICULTURE ISSUES AND POLICIES

METHANE DIGESTERS AND BIOGAS RECOVERY

Agriculture Issues and Policies

Additional books in this series can be found on Nova's website under the Series tab.

Additional E-books in this series can be found on Nova's website under the E-book tab.

Environmental Science, Engineering and Technology

Additional books in this series can be found on Nova's website under the Series tab.

Additional E-books in this series can be found on Nova's website under the E-book tab.

AGRICULTURE ISSUES AND POLICIES

METHANE DIGESTERS AND BIOGAS RECOVERY

JAMES M. HAYWORTH
EDITOR

Nova Science Publishers Inc.
New York

Library of Congress Cataloging-in-Publication Data

Methane digesters and biogas recovery / editors, James M. Hayworth.
p. cm.
Includes index.
ISBN 978-1-61324-594-1 (softcover)
1. Farm manure in methane production. 2. Biomass energy. 3. Manure gases. 4. Methane. 5. Anaerobic bacteria. I. Hayworth, James M.
TP359.M4M485 2011
363.738746--dc23

2011014544

Published by Nova Science Publishers, Inc. † New York

CONTENTS

PREFACE

Biogas recovery systems collect methane from manure and burn it to generate electricity or heat. Burning methane reduces its global warming potential, thereby reducing greenhouse gas (GHG) emissions. Climate change mitigation policies that effectively put a price on GHG emissions could allow livestock producers to "sell" these reductions to other greenhouse gas emitters who face emissions caps or who voluntarily wish to offset their own emissions. Depending on the direction and scope of future climate change legislation, income from carbon offset sales could make methane digesters profitable for many livestock producers. This book examines methane digesters and biogas recovery systems and the economic adoption considerations.

Chapter 1- Biogas recovery systems collect methane from manure and burn it to generate electricity or heat. Burning methane reduces its global warming potential, thereby reducing greenhouse gas (GHG) emissions. Climate change mitigation policies that e ectively put a price on GHG emissions could allow livestock producers to “sell” these reductions to other greenhouse gas emitters who face emissions caps or who voluntarily wish to offset their own emissions. Depending on the direction and scope of future climate change legislation, income from carbon offset sales could make methane digesters profitable for many livestock producers. By modeling the main determinants of producers’ decisions to adopt biogas recovery systems, we illustrate how the price of carbon in uences this decision and the potential supply of carbon offsets from the livestock sector.

Chapter 2- Methane digester systems capture methane from lagoon or pit manure storage facilities and use it as a fuel to generate electricity or heat. In addition to providing a renewable source of energy, digesters can reduce greenhouse gas emissions, odors from manure, and potential contamination of surface water. Methane digesters have not been widely adopted in the United States mainly because the costs of constructing and maintaining these systems have exceeded the value of the benefits provided to the operator. Policies to reduce greenhouse gas emissions could create new opportunities for livestock producers to earn revenue from burning methane from manure, making such biogas recovery facilities profitable for many livestock producers. However, there is likely to be wide variation in the scale, location, and characteristics of livestock operations that would benefit, so these policies could have longrun structural implications for the U.S. livestock sector. In this report we estimate the number and type of hog and dairy operations that would find it profitable to adopt a digester at any given carbon price. We also estimate the relationship between the price of carbon (CO_2) and the amount of emissions reduced by digesters on these operations.

Chapter 3- Livestock producers face a significant challenge in managing manure and process water in a way that controls odors and protects environmental quality. Additionally, livestock manure management practices in the United States are estimated to emit about 2 million tons of methane and account for approximately 8 percent of U.S. methane emissions from anthropogenic activities annually (Figure 1). Biogas recovery systems can provide a multitude of benefits including odor control, improved air and water quality, improved nutrient management flexibility, and the opportunity to reduce greenhouse gas emissions and capture biogas—a useful source of energy.

In: Methane Digesters and Biogas Recovery ISBN: 978-1-61324-594-1
Editor: James M. Hayworth

Chapter 1

CARBON PRICES AND THE ADOPTION OF METHANE DIGESTERS ON DAIRY AND HOG FARMS

United States Department of Agriculture

Biogas recovery systems collect methane from manure and burn it to generate electricity or heat. Burning methane reduces its global warming potential, thereby reducing greenhouse gas (GHG) emissions. Climate change mitigation policies that e ectively put a price on GHG emissions could allow livestock producers to "sell" these reductions to other greenhouse gas emitters who face emissions caps or who voluntarily wish to offset their own emissions. Depending on the direction and scope of future climate change legislation, income from carbon offset sales could make methane digesters profitable for many livestock producers. By modeling the main determinants of producers' decisions to adopt biogas recovery systems, we illustrate how the price of carbon in uences this decision and the potential supply of carbon offsets from the livestock sector.

Methane digesters that collect and burn methane from manure can provide numerous benefits to livestock producers and the environment. They can supply a renewable source of electricity that can power farm equipment or be sold to the electricity distribution grid. Digesters can reduce GHG emissions, odors from manure, and the potential for surface-water contamination. They can also be used to recycle manure solids for animal bedding material. Despite their benefits, digesters have not been widely adopted, mainly because the costs of constructing and maintaining these systems have exceeded the benefits accruing to operators. Of 157 methane digesters operating in the United States as of October 2010, 126 were on dairies and 24 were on hog operations.[1]

Methane is a potent greenhouse gas. Burning 1 ton of methane is equivalent to eliminating about 24 tons of carbon dioxide. A number of policies could encourage farmers to use a digester, either by subsidizing those who install one or penalizing those who do not. While likely impacts on the environment and farm structure vary depending on the policy approach, one of the most promising approaches for controlling GHG emissions is to establish a market price for reductions in GHG emissions.

A carbon offset market is one mechanism for valuing methane emission reductions that is currently in use. An offset market allows livestock producers who reduce methane emissions to sell these reductions or "carbon offsets" to other greenhouse gas emitters (see box, "Carbon Offset Markets"). Currently, only a few U.S. livestock operators sell offsets in regional or voluntary carbon offset markets, partly because carbon prices have been low. However, future e orts to reduce GHG emissions could result in substantially higher carbon prices, which could provide a new source of income for farmers who adopt methane digesters.

Methane Digesters and Greenhouse Gas Emissions

When manure is kept in oxygen-free (i.e., anaerobic) environments like lagoons, ponds, tanks, or pits, it decomposes to produce a biogas containing about 60 percent methane. In contrast, when manure is in oxygen-rich environments, such as when it is deposited on fields, it generally produces little methane, though it can contribute to other environmental problems.

The agriculture sector is responsible for about 6 percent of total U.S. greenhouse gas emissions, of which about 0.6 percent (10.5 percent of agriculture's GHG) is from manure-based methane. Dairy cattle and swine production are responsible for 43 percent and 44 percent of methane emissions from manure, respectively. The other livestock sectors—including beef cattle, sheep, poultry, and horses—are collectively the source of only 13 percent of total manure methane, mainly because manure from these animals is usually handled in ways that produce little methane.

Carbon Offset Markets

In a carbon offset market, farmers are paid for emission reductions without incurring Government expenditures. Farmers sell emission reductions to individuals or firms who wish to "offset" their own emissions. Offsets are measured in tons of carbon dioxide-equivalent emissions (greenhouse gases such as methane are converted to an equivalent quantity of carbon dioxide based on global warming potential). Carbon offsets can be exchanged in markets established to satisfy regulatory compliance or in voluntary markets.

Compliance markets develop when regulations limit the amount of GHGs that firms are allowed to emit, but permit regulated firms to trade emission allowances. Under such a system, known as cap-and-trade, regulated ~ rms (such as power plants) must obtain permits to emit GHGs. To meet their emission targets, firms can reduce their own emissions or purchase permits from other "capped" firms. Or, regulated firms can pay non-regulated emitters, such as livestock operations, to reduce emissions.

Compliance markets currently exist at many levels. International compliance markets include the Kyoto Protocol and the European Union's Emissions Trading Scheme. Ten Eastern States recently implemented the Regional Greenhouse Gas Initiative (RGGI), the first mandatory market-based e ort in the United States to reduce GHG emissions. The Federal Government recently considered climate change legislation that would establish a national cap-and-trade system.

Recommended Citation Format for This Publication

Key, Nigel, and Stacy Sneeringer. Carbon Prices and the Adoption of Methane Digesters on Dairy and

Hog Farms, EB-16 U.S. Department of Agriculture, Economic Research Service, February 2011.

A biogas recovery system--known variously as an "anaerobic digester," "methane digester," "biodigester," or "methane recovery system--"captures methane from anaerobic manure storage facilities. Such systems collect manure, optimize it for the production of methane by adjusting temperature and water content, capture the biogas, and burn it for heat or electricity generation.

Voluntary offset markets allow companies and individuals to voluntarily purchase carbon offsets. For example, individuals might seek to offset their travel-related emissions or firms might seek to compensate for emissions related to their products. In the United States, the Chicago Climate Exchange (CCX) is a voluntary, but legally binding, carbon trading regime.

FACTORS INFLUENCING METHANE DIGESTER ADOPTION

A farmer's decision to adopt a methane digester depends on the relative costs and benefits of doing so. Costs include start-up and maintenance expenditures. Benefits include revenues from the sale of electricity, foregone electricity purchases, revenue from the sale of carbon offsets, and less odor from stored manure. A number of factors determine these costs and benefits, including the size of the operation, manure storage method, the price of electricity, onfarm electricity expenditures, the ability to sell electricity not used on the farm, and—if there is a carbon offset market—the price of carbon. A model of digester profitability illustrates how these factors a ect the number of farms that might adopt digesters (see box, "Modeling Methane Digester Profits"), how the price of carbon offsets would in uence the quantity of methane emitted from manure management, and how large the supply of carbon offsets from digester adoption might be.

Manure Management Method and Farm Size

Potential revenue from a digester system depends on the type of manure storage facility in use. Offset programs usually require documentation of baseline emissions and certification that offsets lead to "additional" emission reductions. Consequently, only livestock operations using anaerobic manure storage before the creation of an offset market would likely qualify for an offset program. This limits the pool of potential offset market participants to swine and dairy operations having manure ponds, lagoons, or slurry pit systems. Operations with slab or shed manure systems or with no storage facilities would not generate sufficient methane to satisfy the "additionality" requirements for offset certification. We estimate that up to 42 percent of dairies and 64 percent of hog operations have manure management systems that qualify for an offset program.

Anaerobic digesters are generally added to two main types of manure storage. These are "lagoon" systems in which a cover is placed over an earthen storage pond, and "pit" systems in which manure is processed through a heated tank to induce methane production. Lagoon systems are less expensive to construct, but often produce less electricity in cooler climates because they are usually not heated. Pit-based systems are more expensive to build, but can produce more electricity when heated.

The type of manure storage facility used determines the baseline methane emitted and, consequently, the quantity of carbon offsets that could be generated and sold. Lagoon systems generally emit higher rates of methane per head than pit systems, and operations in warmer climates emit more than those in cooler climates. While pit systems in cooler climates can be heated to generate more methane (and consequently more electricity), this extra methane would not be considered in the baseline emissions, and so would likely not qualify for carbon offsets.

Figure 1 illustrates how profits from a biogas recovery system vary across manure management systems in the two biggest dairy States, Wisconsin and California. The figure compares the net present value (NPV) per head of a digester project that lasts 15 years when the carbon offset price is $13 (per ton of carbon dioxide-equivalent emissions). In both States, operations with lagoons are more profitable than operations with pit systems, mainly because lagoons have much higher initial emissions and consequently higher offset revenues. Also, biogas systems are more profitable in California than in Wisconsin. California's warmer climate increases methane production and lowers operating and construction costs, and electricity prices are higher there than in Wisconsin.

Farm size is another important determinant of digester profits. Digester construction at dairy operations with 500-1,000 head is estimated to cost between $366,000 and $652,000, while annual maintenance costs are estimated to range between $7,000 and $28,000. Construction costs per head of livestock generally decline as the operation increases in size, making adoption more profitable for larger operations. A digester on a 1,500-head Wisconsin dairy with a lagoon manure system (purple line) has a NPV of $333 per head. In contrast, a 1,000-head dairy in Wisconsin using the same manure management system would have a NPV of only $239 per head (Figure 1). The additional income available to larger livestock operations from biogas recovery systems could enhance their competitive advantage over smaller producers, and could contribute to the ongoing concentration of production on larger farms.

Modeling Methane Digester Profits

An investment model is used to estimate how farm size, manure management method, and the carbon off set price aff ect a producer's decision to adopt a biogas recovery system. A hog or dairy producer is assumed to adopt a digester if the net present value (NPV) of the project is positive. The NPV is the sum over the life of the project (15 years) of all cash flows (e.g., revenues from electricity or carbon off sets minus capital and variable costs), discounted to the present value:

NPV = Present value of revenues – Present value of costs.

The gross revenue from the digester includes electricity expenditures avoided by generating one's own electricity (and not having to buy it from the utility company), sales of electricity not used onfarm, and sales of carbon off sets:

Revenues = Electricity expenditures avoided + Electricity sales + Carbon offset sales.

The amount of electricity expenditures avoided depends on the amount of electricity used onfarm and the price of purchased electricity. The revenue from electricity sold depends on the excess amount generated and the price received for electricity sold. The price for electricity purchased may be diff erent from the price of electricity sold, though they are set equal in this analysis. The price of electricity is modeled to increase with the price of carbon.

The revenue from carbon offsets will depend on the amount of methane generated and the price of carbon. The amount of methane generated depends on the operation's manure management system and the amount of manure produced, which depends on farm size. Other benefits from a digester—like odor reduction, water pollution abatement, and revenues from separated solids and from "tipping fees" (fees paid to deposit food wastes into the digester)—are not included in the model because it is di cult to assess their value. Consequently, the model predicts a lower bound to the total benefits from a biogas system.

The costs of the digester include construction (capital, equipment, etc.) costs, maintenance and repair costs, and transaction costs associated with participating in a carbon offset market:

Costs = Construction costs + Maintenance costs + Offset market costs.

The model assumes that an operation will pay off the initial construction costs of the digester over a 15-year period.

The quantity of electricity that can be generated and the construction and maintenance costs of the digesters are functions of farm size and type of manure handling system. These functions are specified using information from case studies. State-level electricity price data are from the U.S. Department of Energy. Onfarm electricity use is observed in the Agricultural Resource Management Survey (ARMS) 2005 dairy survey and 2004 hog survey. Methane emissions are estimated using State-level coefficients based on Intergovernmental Panel on Climate Change methodologies. This analysis assumes no financial support or other programs to subsidize methane digester costs.

The model is used to estimate digester profits for every farm in the ARMS 2005 dairy and 2004 hog surveys. Survey weights are used to develop estimates of digester adoption rates and the carbon offset supply at the national level.

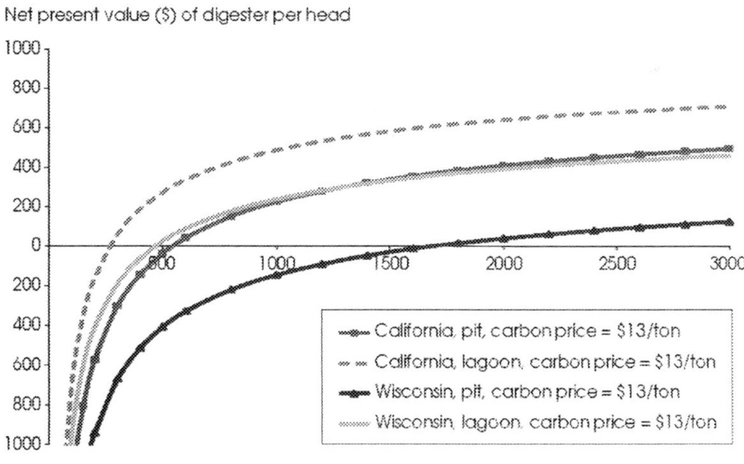

Source: Authors' calculations, using 2005 ARMS data.

Figure 1. Net present value per head for methane digesters on California and Wisconsin dairies using lagoon and pit manure management systems.

Electricity Prices, Onfarm Electricity Use, and Sales of Excess Electricity

Revenues from the sale of surplus electricity and cost savings from farm-generated electricity depend on the price of electricity, which varies substantially across States. The electricity generated in 1 year by a 1,000-head dairy with a pit-based digester would be worth approximately $56,300 (retail) in Wisconsin, versus $77,500 in California. If farmers generate more electricity than they use and they are able to sell this surplus electricity, then a higher electricity price increases their incentive to adopt a methane digester.

If operations are unable to sell surplus electricity back to the electricity distribution grid, then the benefits from electricity generation are limited to the avoidance of onfarm energy costs associated with heating or cooling, drying grain, pumping water, lighting, and operating machinery. Onfarm energy expenditures vary widely across regions. The 2005 Agricultural Resource Management Survey (ARMS) dairy survey indicates that a 1,000-head dairy in Wisconsin typically spends about $125,700 per year on energy (electricity, natural gas, and propane; amount updated to 2009 dollars), more than double what it could generate onfarm with a pit-based digester ($56,300). In contrast, a 1,000-head California dairy typically spends about $53,600 on energy (2009 dollars), so it would use less energy than it could generate ($77,500)[2] Consequently, without the ability to sell electricity, farms in California would receive only a fraction of their generated electricity's potential value. In this example, the Wisconsin farm with higher energy use (due to higher heating costs) would have a greater incentive to adopt a biogas recovery system than the California farm, despite having lower electricity prices. However, if the California farm could sell its surplus electricity at a sufficiently high price, then it could benefit more from adopting a digester than the Wisconsin farm because the California farm can generate more electricity.

One potential problem with using electricity from biogas facilities is that the onfarm supply may not match demand. The quantity of electricity generated may fluctuate over the day, month, or year depending on temperature, inflows of manure, machine malfunctions, and the like. Similarly, onfarm electricity use fluctuates over time. "Net metering" laws mitigate this problem by allowing small-scale generators to obtain the full retail value for the electricity they generate. Under net metering laws, when surplus electricity is produced onfarm, the electricity meter spins backwards, eff ectively saving the electricity until it is needed. Over the billing period, the operation is only billed for its net electricity usage. More than 40 States have net metering laws.

Operations that generate more electricity than they consume over a billing period may be able to sell their excess electricity to the utility at a negotiated price. The ability to sell surplus electricity and the selling price vary regionally. Recent laws and trends suggest that an increasing number of livestock operations may be able to sell electricity at retail or even higher prices. Since manure-derived electricity is from a renewable source, the negotiated price for excess electricity could enjoy a substantial premium over the wholesale price. Currently, about 30 States require utilities to purchase a share of power from renewable sources, including farms with biogas systems.

Carbon Offset Price

The additional revenues that could be earned from carbon off sets could have a large eff ect on digester profitability and adoption if the offset price is sufficiently high. However, future carbon prices are uncertain. In the major international compliance markets, carbon prices have ranged between $15 and $30 per ton of carbon dioxide-equivalent emissions in the last decade. In the United States, off set prices have been much lower. The average price for carbon allowances in the eastern States has ranged between $1 and $3 per ton since the RGGI's inception in 2008. The CCX carbon price has ranged between $1 and $7 per ton since 2004, but has been trading under $1 per ton since 2009. The Environmental Protection Agency estimates that the recently proposed House bill (H.R. 2454, the "American Clean Energy and Security Act of 2009") would have resulted in a carbon off set price of $13 per ton had it been enacted.

At a particular carbon price, some farmers will choose to adopt digesters based on their farm's size, location, manure storage system, electricity prices, and onfarm electricity use. A higher carbon price increases the potential revenue from off set sales. A higher carbon price would also likely be associated with higher electricity costs, which would increase the value of electricity generated with a methane digester.

As carbon prices increase, more livestock operations would find it profi table to adopt a digester. Figure 2 illustrates, at three diff erent carbon prices, the estimated number of dairy operations in diff erent size categories on which a digester would have a positive NPV. With a carbon price of $0 (no off set market), digesters would have a positive NPV on only 69 operations with at least 1,000 head and 2 operations with fewer than 1,000 head. However, if the off set price were $13 per ton of carbon dioxide equivalent, then digesters would have a positive NPV on 658 operations with at least 1,000 head and on 1,190 operations with fewer than 1,000 head. If the price were to increase to $26, digesters would have a positive NPV on many more small-scale operations, including 3,575 operations with fewer than 1,000 head.

Higher carbon prices cause more farms to find methane digesters profi table. If these farms install a digester, then the total greenhouse gas emissions from manure management would fall. By predicting which operations would adopt a digester (i.e., those on which a digester has a positive NPV) and then summing the reduction in tons of carbon dioxide-equivalent emissions, it is possible to generate a curve representing the relationship between the price of carbon and the amount of emissions reduced.

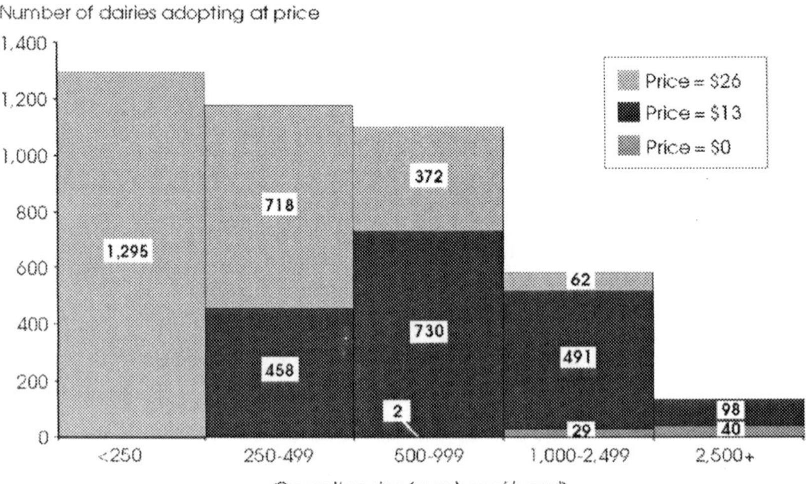

Notes: Numbers at higher prices are additive to those for lower prices; for example, at a price of $13/ton, an additional 491 operations of size 1,000-2,499 head are predicted to adopt, for a total of 520 operations of this size. At a carbon price of $13/ton, no operation smaller than 250 head is predicted to adopt. At a carbon price of $0, no operation with fewer than 500 head and 2 operations 500-999 head are predicted to adopt.

Figure 2. Number of dairy operations on which digesters have a positive net present value at different carbon offset prices.

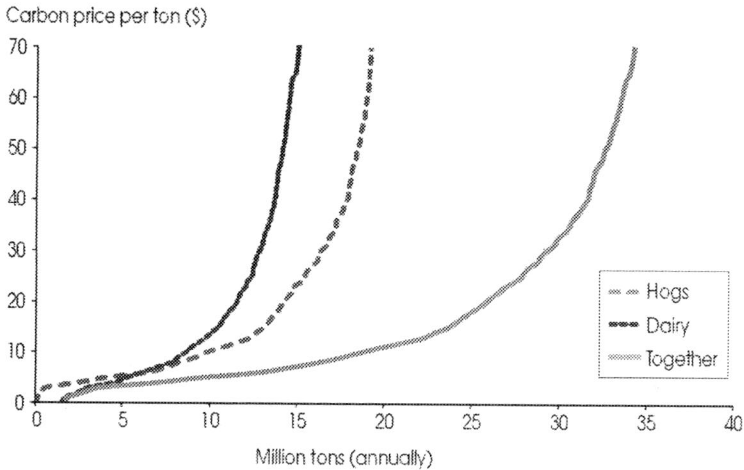

Source: Authors' calculations, using 2004 and 2005 ARMS data.

Figure 3. Total reductions in manure methane emissions from dairy, hogs, and both sectors together at different carbon offset prices.

Figure 3 shows this relationship for dairies, hog operations, and both livestock types combined. Without a carbon market (when the price is zero), no hog operations find it profitable to adopt a digester, so there is essentially no reduction in emissions. In contrast, some dairies find that electricity generation alone would make adoption profitable, so some emissions are reduced when the carbon price is zero. As the carbon price increases, more operations adopt digesters, lowering emissions. Eventually, at a price of about $70, the curves approach vertical and almost all of the total potential reduction of methane is reached.

At a carbon price of $13, greenhouse gas emissions are reduced by 9.8 and 12.4 million tons (carbondioxide equivalent) for the dairy and hog sectors, respectively. This amounts to reductions of 61-62 percent of manure-generated methane in these sectors. A doubling of the carbon price to $26 would cause manure-based methane emissions from dairy and hogs together to be reduced by 78 percent. These emission reductions represent the potential supply of carbon off sets from hog and dairy producers who could earn profits by adopting a methane digester. The actual off set supply will depend on how many of these farmers choose to adopt a digester and participate in an off set market.

At an off set price of $13, the net present value (i.e., the present value of the stream of net revenue earned over 15 years) of all digesters in the dairy and hog sectors together would be about $1.8 billion. At this carbon price, off set fees would comprise the largest source of digester revenue (66 percent of total), far exceeding savings in electricity costs (26 percent) and surplus electricity sales (8 percent).

CONCLUSIONS

The extent to which livestock operations can reduce greenhouse gas emissions from manure management and the potential supply of carbon off sets from the livestock sector depend fundamentally on the number of livestock operations that adopt a methane digester. The adoption decision depends on digester profitability—which in turn depends on the value of the electricity generated, revenues from the sales of carbon off sets, and other benefits and costs of the facility. Key factors aff ecting digester profitability include:

- ***The size of the operation.*** Larger operations are more likely to find participating in offset markets profitable because of economies of scale in digester technology.
- ***Farm electricity expenditures***, which depend on electricity prices and onfarm energy consumption. Greater onfarm expenditures make digesters more profitable if an operation is unable to sell its surplus electricity or if the selling price is low.
- ***The selling price of surplus electricity***. A higher price makes digester-generated electricity more valuable for operations that can generate more electricity than they use onfarm.
- ***Initial levels of methane emissions***. The amount of methane generated before adopting a digester determines the quantity of "additional" emission reductions that can be sold as offsets. Farms with lagoons located in warmer regions generally produce a greater initial quantity of emissions than those with pit manure systems located in cooler climates.
- ***The carbon offset price***. A higher carbon price means greater revenues from offset sales.

ERS research indicates that revenues from the sale of carbon off sets could have a substantial eff ect on the number of operations that would adopt a biogas recovery system. A carbon price of $13 could induce dairy and hog operations to supply off sets equivalent to about 22 million tons of carbon dioxide annually. This amounts to 62 percent of the potential off sets from manure management in these sectors, or about 5 percent of total greenhouse gas emissions from U.S. agriculture.

Whether or not a carbon off set market is created, several policy tools could increase digester profits and result in more widespread adoption of the technology. For example, further expanding the scope of "net metering" laws and raising minimum renewable energy requirements for utilities would increase the price farmers receive for their generated electricity. Subsidies or tax breaks could directly lower digester construction costs, while public research to reduce construction and maintenance costs and improve digester efficiency could raise the economic feasibility of digesters over the longer term.

This brief is drawn from: Climate Change Policy and the Adoption of Methane Digesters on Livestock Operations. Economic Research Report 111. It is the third in a continuing series of ERS economic briefs examining agriculture's potential role in climate mitigation and the economic and policy issues involved.

The U.S. Department of Agriculture (USDA) prohibits discrimination in all its programs and activities on the basis of race, color, national origin, age, disability, and, where applicable, sex, marital status, familial status, parental status, religion, sexual orientation, genetic information, political beliefs, reprisal, or because all or a part of an individual's income is derived from any public assistance program. (Not all prohibited bases apply to all programs.) Persons with disabilities who require alternative means for communication of program information (Braille, large print, audiotape, etc.) should contact USDA's TARGET Center at (202) 720-2600 (voice and TDD).

To file a complaint of discrimination write to USDA, Director, Office of Civil Rights, 1400 Independence Avenue, S.W., Washington, D.C. 20250-9410 or call (800) 795-3272 (voice) or (202) 720-6382 (TDD). USDA is an equal opportunity provider and employer.

End Notes

[13] U.S. Environmental Protection Agency, *U.S. Anaerobic Digester Status Report*, October 2010.

[2] These amounts were calculated using 2009 electricity prices for the industrial sector; these prices were $0.0921/ kwh for California and $0.067/ kwh for Wisconsin. Source: U.S. Department of Energy, U.S. Energy Information Administration, Office of Coal, Nuclear, and Alternate Fuels. 2010. *Electric Power Monthly*. June 2010. DOE/IEA-0226. Table 5.6.B. Accessed at: *http://www.eia.doe.gov/cneaf/electricity/epm/matrix96_2000. html/*.

In: Methane Digesters and Biogas Recovery ISBN: 978-1-61324-594-1
Editors: James M. Hayworth

Chapter 2

CLIMATE CHANGE POLICY AND THE ADOPTION OF METHANE DIGESTERS ON LIVESTOCK OPERATIONS

Nigel Key[*] *and Stacy Sneeringer*[**]

ABSTRACT

Methane digesters—biogas recovery systems that use methane from manure to generate electricity—have not been widely adopted in the United States because costs have exceeded benefits to operators. Burning methane in a digester reduces greenhouse gas emissions from manure management. A policy or program that pays producers for these emission reductions—through a carbon offset market or directly with payments—could increase the number of livestock producers who would profit from adopting a methane digester. We developed an economic model that illustrates how dairy and hog operation size, location, and manure management methods, along with electricity and carbon prices, could influence methane digester profits. The model shows that a relatively moderate increase in the price of carbon could induce significantly more dairy and hog operations, particularly large ones, to adopt a methane digester, thereby substantially lowering emissions of greenhouse gases.

[*] nkey@ers.usda.gov
[**] ssneeringer@ers.usda.gov

Keywords: methane, methane digesters, manure, livestock, climate change, greenhouse gases, carbon offset

ACKNOWLEDGMENTS

The authors would like to thank Brent Gloy, Department of Agricultural Economics, Purdue University; Jeffrey Hyde, Department of Agricultural Economics and Rural Sociology, Pennsylvania State University; John Horowitz, USDA, Economic Research Service; Richard Hegg and Luis Tupas, USDA, National Institute of Food and Agriculture; and Robert Johansson and Jan Lewandrowski, USDA, Office of the Chief Economist, for their helpful reviews and comments. Thanks also go to our editor, Priscilla Smith, and our designer, Wynnice Pointer-Napper.

SUMMARY

What Is the Issue?

Methane digester systems capture methane from lagoon or pit manure storage facilities and use it as a fuel to generate electricity or heat. In addition to providing a renewable source of energy, digesters can reduce greenhouse gas emissions, odors from manure, and potential contamination of surface water. Methane digesters have not been widely adopted in the United States mainly because the costs of constructing and maintaining these systems have exceeded the value of the benefits provided to the operator. Policies to reduce greenhouse gas emissions could create new opportunities for livestock producers to earn revenue from burning methane from manure, making such biogas recovery facilities profitable for many livestock producers. However, there is likely to be wide variation in the scale, location, and characteristics of livestock operations that would benefit, so these policies could have longrun structural implications for the U.S. livestock sector. In this report we estimate the number and type of hog and dairy operations that would find it profitable to adopt a digester at any given carbon price. We also estimate the relationship between the price of carbon (CO_2) and the amount of emissions reduced by digesters on these operations.

What Are the Major Findings?

The extent to which livestock operations can reduce greenhouse gas emissions from manure management depends in part on the number of livestock operations that adopt methane digesters, which in turn depends on digester profitability from energy savings, energy sales, and/or sales of emission reductions in a carbon offset market. An offset market allows livestock producers who reduce methane emissions to sell these reductions or "carbon offsets" to other greenhouse gas emitters who might face emissions caps.

Factors that influence digester profitability and that determine the characteristics and locations of the livestock operations that could benefit from the introduction of a carbon offset market include:

- operation size—costs of constructing and operating a digester decline on a per-head basis, making digesters more profitable on larger operations
- the selling price of surplus electricity—a higher price makes digesters more valuable for operations that can generate more electricity than they use onfarm
- farm electricity expenditures, which depend on electricity prices and onfarm use—higher expenditures make digester-generated electricity more valuable, especially if the operation cannot sell electricity or if the selling price of electricity is below the retail price
- participation in cost-share and other incentive programs—this can defray the cost of building digesters
- farm's initial level of methane emissions—this determines the maximum quantity of carbon emissions reductions that can be sold
- carbon price—a higher carbon price makes digesters more profitable for operations that can sell carbon offsets.

Larger operations would be more likely to adopt a digester, and likely would earn substantially higher profits on average than smaller operations. Hence, introduction of a carbon market in a region could enhance existing economies of scale in production and result in further concentration of production on the largest operations. However, smaller livestock operations may be able to achieve a more efficient digester scale by supplementing manure with food waste products or by sharing a digester with other small operations. In addition, if the adoption of methane digesters by smaller operations is a policy goal, several tools exist—such as cost-share subsidies or tax incentives—that could be used to encourage their adoption by small farms.

Additional revenues from the sale of carbon emissions reductions (offsets) could substantially increase the number of operations that would adopt a biogas recovery system. Findings in this study indicate that a carbon price of \$13 per metric ton of carbon dioxide equivalent emissions (an initial price estimated under one scenario for a nationwide cap-and-trade program for greenhouse gases) would:

- induce dairy and hog operations to supply offsets equivalent to about 22 million tons of carbon dioxide annually, amounting to about 62 percent of the current greenhouse gas emissions from manure management in these industries, or about 5 percent of total greenhouse gas emissions from the U.S. agricultural sector
- allow dairy and hog operators as a group to earn up to \$1.8 billion in additional profits over 15 years from installing methane digesters.

Currently, the price of electricity and onfarm electricity expenditures are key determinants of digester profitability. However, when carbon prices are above \$4 per metric ton of CO_2 equivalent emissions, carbon offset sales comprise a larger source of digester revenue than electricity generation. At a price of \$13 per metric ton of CO_2 equivalent emissions, revenues from emission reduction sales (offsets) contribute 66 percent of gross digester revenues for all dairy and hog operations, electricity sales contribute 8 percent, and cost savings from avoided energy expenses contribute the remaining 26 percent.

At higher carbon prices, the distribution of profits from digesters refl ects the location of large-scale operations and the prevalence of lagoons. Among States with the greatest number of dairies, the study finds that California, New York, Wisconsin, and Texas each have at least 100 such operations that would find it profitable to adopt a digester at a carbon price of \$13 per metric ton of CO_2 equivalent emissions. At the same price, North Carolina, Illinois, Indiana, Missouri and Oklahoma each have at least 100 hog farm operators who would find a methane digester profitable.

How Was the Study Conducted?

We used a model of digester profitability to estimate how farm size, manure management methods, electricity prices, and carbon prices affect producers' decisions to adopt biogas recovery systems. Hog and dairy producers are assumed to adopt a digester if the present value of the discounted stream of profits (the net present value) is positive. Profits derive from electricity generation and carbon emission reductions sales less the digester construction and maintenance costs. Using case study information, we parameterized the model. Electricity price data are drawn from the U.S. Department of Energy, and methane emissions are estimated using State-level Intergovernmental Panel on Climate Change emission coefficients.

By computing the present value of digester profits for every farm in nationally representative samples of dairy and hog operations (USDA's Agricultural Resource Management Survey or ARMS), we used the model to provide an estimate of the number, size, and location of farms that would find it profitable to adopt a digester at any given carbon price. ARMS is conducted by USDA's National Agricultural Statistics Service (NASS) in conjunction with the Economic Research Service. By predicting which operations would earn profits from digester adoption and then summing the reduction in tons of carbon dioxide equivalent emissions, it is possible to estimate the relationship between the price of carbon and the amount of emissions reduced by methane digesters on dairy and hog operations. We used the model to estimate how the present value of farm revenues changes with the carbon price and to simulate the effect of surplus electricity prices and Government cost-share policies on the potential supply of carbon emissions reductions.

INTRODUCTION

Methane digesters, also known as "anaerobic digesters," "biodigesters," or "biogas recovery systems," can be used to capture and burn methane from lagoon or pit manure-storage facilities. With lagoons (ponds surrounded by earthen berms), covers are installed to capture the methane. With pits (concrete or metal tanks located above or below ground), manure can be more easily heated to maximize methane production. Manure is collected and transported to the digester, where the water, nutrients, and heat are adjusted to optimize methane output. Digesters capture the biogas, treat it, and send it to a boiler or electricity generator.

Methane digesters can provide numerous benefits to livestock producers and the environment. Digesters can supply a renewable source of electricity that can be used on the farm or sold to the electricity distribution grid. Digesters can reduce greenhouse gas emissions, odors from manure, and the potential for surface water contamination. They can also be used to recycle manure solids for animal bedding material. Despite their benefits, digesters have not been widely adopted, mainly because the costs of constructing and maintaining these systems have exceeded the benefits accruing to operators. Currently, there are 157 methane digesters operating in the United States, of which 126 are on dairies and 24 are on hog operations (U.S. EPA, 2010a).

Digesters have received more attention lately because of their ability to reduce greenhouse gas emissions from livestock manure. Lagoon and pit manure handling systems that are common on dairy and hog operations emit large amounts of methane. Methane is a potent greenhouse gas (GHG). Burning 1 ton of methane is equivalent to eliminating about 24 tons of carbon dioxide (IPCC, 2007, table 2.14).[1] There are a number of policies that could provide financial incentives for farmers to use a digester to reduce methane emissions, and the likely impacts on the environment and farm structure vary depending on the policy approach. One approach for controlling GHG emissions that requires relatively little direct Government financial support is to establish a market for GHG emissions reductions, or "carbon offsets."

An offset market allows livestock producers who reduce methane emissions to sell these reductions to other greenhouse gas emitters who face emissions caps or who voluntarily wish to offset their own emissions. Currently, only a few U.S. livestock operators sell offsets in regional or voluntary carbon markets. This is partly because the carbon prices in these markets have been low. Future efforts to reduce greenhouse gas emissions could result in substantially higher carbon prices. Prospects for a national carbon offset market within a cap-and-trade framework are uncertain. However, 10 Eastern States currently have a regional offset market (Regional Greenhouse Gas Initiative) and 7 other States are developing a regional cap-and-trade system (Western Climate Initiative).

Farmers who adopt methane digesters could profit from higher carbon prices. However, there is likely to be wide variation in the scale, location, and characteristics of the farm operations that would benefit from having methane digesters. The main beneficiaries would be producers whose farm operations emit substantial quantities of methane, particularly dairy and hog operations with lagoon or pit manure storage facilities. Among these, larger scale operations likely would profit more from higher carbon prices because constructing and operating larger digesters is generally more cost-effective. As a result, valuing emission reductions could have significant effects on the long-term structure of the livestock industry. Smaller operations may be shut out of the profits of an expanded carbon offset market unless ways are found to promote the adoption of digesters on small-scale operations.

In addition to these structural implications, understanding the extent to which the agricultural sector, and livestock in particular, participates in offset markets is important because of the effect this will have on "capped" industrial sectors and on agricultural producers. The supply of offsets will help determine the price of carbon emission permits and hence the costs that the "capped" industries face in meeting their emissions permit requirements. For livestock producers, the offset revenues could provide an additional source of income that could compensate them for higher feed or energy costs resulting from climate legislation. In this report, we explore how the carbon price could influence the supply of carbon offsets from the livestock sector. We also estimate the level and geographical distribution of income from carbon offset sales at different carbon prices.

DAIRY, HOG OPERATIONS PRODUCE MOST MANURE METHANE

Livestock generate large amounts of manure that must be stored, spread on fields, or moved off-farm. Manure mixed with water is often stored in lagoons, ponds, tanks, or pits, creating anaerobic (i.e., without oxygen) conditions. The decomposition of livestock or poultry manure without oxygen produces a biogas containing about 60 percent methane.[2] When manure is handled as a solid or deposited on fields, it tends to decompose aerobically (i.e., with oxygen) and produces much less methane. The quantity of methane released also depends on climate (temperature and rainfall) and the conditions under which manure is managed (oxygen level, water content, pH (acidic or basic qualities) level, and nutrient availability).

In 2008, the U.S. agricultural sector was responsible for 6.1 percent of total U.S. greenhouse gas emissions (U.S. EPA, 2010b, p 2-12).[3] Methane emissions from manure management were responsible for about 10.5 percent of these agricultural emissions.[4] Dairy cattle and swine producers, who often use anaerobic manure management systems, were responsible for 43.1 percent and 43.6 percent of methane emissions from manure management, respectively (U.S. EPA, 2010b, table 6-6).[5] Beef cattle, sheep, poultry and horses were collectively the source of only 13.3 percent of total manure methane, mainly because manure from these animals is usually handled in aerobic conditions.[6] Geographic shifts and increasing scale of production have led to a greater share of dairy cattle and swine being raised in facilities using anaerobic storage facilities. This, in turn, has resulted in a substantial increase in methane emissions from manure handling—emissions increased 54 percent between 1990 and 2008 (U.S. EPA, 2010b, table 6-2). By trapping and burning methane, digesters have the potential to substantially reduce greenhouse gas emissions from manure management.[7]

Climate Change Policies Could Spur Digester Adoption

There are several possible policy approaches to mitigate greenhouse gases, and each could have different implications for livestock producers and their decisions to adopt methane digesters. One approach is to place controls on individual emitters by regulating the production technologies that can be used. Livestock producers, as emitters of greenhouse gases, might be required to adopt specific technologies that would reduce emissions, such as lagoon covers that collect and burn methane. Even if livestock producers were not subject to such "technology standards," it is possible that utility companies would be, resulting in higher energy costs being passed along to farmers. Such a rise in energy costs would make digester-generated electricity more valuable to farmers, who would save more by generating their own electricity.

Another approach is to encourage adoption of mitigation technologies through subsidies or other incentives. In the case of digesters, incentives could take the form of grants, cost shares, incentive payments, and State or Federal tax credits or exemptions. As discussed later in this paper, there are several existing programs providing incentives for livestock producers to adopt methane digesters. Many existing incentive programs are designed to promote renewable energy, in addition to lowering GHG emissions.[8]

A third policy approach is to tax greenhouse gas emissions directly or to tax them indirectly by taxing commodities based on their "carbon content."[9] Emissions from livestock operations or livestock products such as dairy products and meat could, in theory, face such a carbon tax. Even if livestock producers or products were not taxed, electricity and petroleum-based fuels could be taxed, which would raise energy costs for livestock producers. Again, this would make digester-generated electricity more valuable, and consequently would encourage digester adoption.

A fourth approach is to pay farmers for emissions reductions. Farmers could be compensated with Government payments or carbon offset sales. In an offset market, farmers sell emissions reductions to individuals or firms who wish to "offset" their own emissions. To be eligible as a carbon offset, emissions reductions generally must meet several criteria, including:

1. additional to "business as usual"—the reductions would not have occurred without the offset sales
2. accurate—the reductions must be verifiable by a third party
3. permanent—the emissions reductions from the offset project are not reversible.

Emissions reductions resulting from the burning of manure methane can satisfy all of these requirements. Offsets are measured in tons of carbon dioxide equivalent emissions (reductions in other greenhouse gases such as methane are converted to an equivalent quantity of carbon dioxide based on their global warming potential).

Carbon offsets can be exchanged in compliance or voluntary markets. Compliance markets usually operate in conjunction with a cap-and-trade regime that places a legal limit on the quantity of greenhouse gases that can be emitted by regulated firms in a particular time period. Under such a system, regulated firms must obtain permits to emit greenhouse gases. To meet their emissions targets, regulated firms can reduce their own emissions or purchase permits from other "capped" firms. Alternatively, when regulations permit, firms could pay nonregulated emitters—such as livestock operations—to reduce emissions (i.e., the firms could purchase offsets).

Compliance markets have been established at the international, national, and regional levels. Regimes that govern international compliance markets include the Kyoto Protocol and the European Union's Emissions Trading Scheme. In the United States, 10 eastern States recently implemented the Regional Greenhouse Gas Initiative (RGGI), the first domestic mandatory market-based effort to reduce greenhouse gas emissions. Under the RGGI, the capped sector (power generation) can purchase emission offsets from other sectors, such as agriculture, that reduce or sequester greenhouse gas emissions. Projects that reduce methane emissions from manure management are eligible for offset allowances. In 2009, the U.S. House of Representatives approved climate change legislation (H.R. 2454) that, if signed into law, would have established a national cap-and-trade system and provided an opportunity for farmers to sell offsets from reducing their manure methane emissions.

Voluntary offset markets function outside of compliance markets and allow companies and individuals to voluntarily purchase carbon offsets. For example, individuals might seek to offset their travel-related emissions or "green" firms might seek to compensate for emissions related to their production. In the United States, the Chicago Climate Exchange (CCX) is a voluntary, but legally binding, carbon trading regime. In this privately administered cap-and-trade system, methane emissions reductions from livestock operations can qualify as offset projects.

Depending on the price of carbon, the additional income from offset sales could substantially increase the number of livestock producers who would find it profitable to install methane digesters.

Boosting Benefits over Costs Would Encourage Methane Digester Adoption

The decision to adopt a digester depends on the price of electricity, onfarm electricity expenditures, the ability to sell electricity not used on the farm, the manure management method employed, the size of the operation, the startup and ongoing costs of the technology, and the value of other benefits such as odor reduction, reduced risk of water and air pollution, and the sale of separated digester solids. Government policies, such as digester cost-share programs or policies that create a demand for electricity from renewable sources, can affect the economic feasibility of methane digesters. The price of carbon in a carbon offset market could play a role in a digester adoption decision.

We developed a farm-level investment model to estimate how many of these multiple factors influence methane digester profits and adoption, and what the supply of carbon offsets would be from the livestock sector. In the model, profits from a digester are equal to the returns from electricity generation and the sales of carbon offsets less the costs associated with constructing and maintaining the digester. Hog and dairy producers are assumed to adopt a digester and sell carbon offsets if doing so has a positive net present value (NPV). The NPV is the sum of all cash flows (e.g., revenues from electricity and carbon offsets minus capital and variable costs) over the

life of the project (assumed to be 15 years), where the cash flow in each year is discounted to its present value.

The model is parameterized using cost and production information derived from several recent case studies of digesters installed on dairy and hog operations. Other livestock sectors were not considered because of a lack o data and the relatively limited contribution of other sectors to total manur methane emissions (U.S. EPA, 2010b). Electricity price data are drawn fron the U.S. Department of Energy, and methane emissions are estimated usin State-level emission coefficients based on Intergovernmental Panel on Climat Change (IPCC) methods.

For a livestock operation of a given size, type and location, the mod provides an estimate of the NPV of the digester. By estimating NPV for eve farm in nationally representative samples of dairy (2005) and hog operatio (2004) (using USDA's Agricultural Resource Management Survey (ARMS the model predicts which farms would adopt a digester at any given carb offset price. ARMS is conducted by USDA's National Agricultural Statist Service in conjunction with the Economic Research Service. By summing reduction in tons of carbon dioxide equivalent emissions for farms that wo adopt a digester (i.e., farms on which a digester has a positive NPV), we estimate the relationship between the price of carbon and the total level emissions reduced by methane digesters on dairy and hog operations. ARMS survey weights allow us to extrapolate these estimates to the natic level. The model also is used to estimate how farm revenues change with carbon offset price and to simulate the effect of surplus electricity prices Government cost-share policies on the potential supply of carbon off More detailed information about the model specification, case studies, and parameters is given in the appendix (p. 32).

Electricity—Prices, Onfarm Use, and Sales of Surplus—Is a to Digester Profitability

The price of electricity is a key factor determining methane di profitability because the price determines the cost savings from generated electricity and the revenues that can be earned from the s surplus electricity. Electricity prices vary substantially across the c (table 1). For example, because of different retail electricity price electricity generated in 1 year by a 1,000-head dairy with a pit-based di would be worth approximately $56,300 (retail) in Wisconsin compare

$77,500 in California. The amounts were calculated using 2009 electricity prices for the industrial sector—$0.0921 per kilowatt hour for California and $ 0.067 per kilowatt hour for Wisconsin (U.S. DOE, 2010). If farms are able to sell surplus electricity to the grid, then the higher electricity prices provide operators with a greater incentive to adopt biogas collectors.

If operations are unable to sell surplus electricity back to the grid, then the benefits from electricity generation are limited to the avoided onfarm energy costs associated with heating or cooling, drying grain, pumping water, lighting, and operating dairy or other machinery. Onfarm energy expenditures per head also vary widely across regions because of differences in climate and production technologies (see table 1). Data from the 2005 ARMS dairy survey indicate that a 1,000-head dairy in Wisconsin typically spends about $125,700 per year on energy (electricity, natural gas, and propane; amount updated to 2009 dollars), which exceeds what it could generate onfarm ($56,300). In contrast, a 1,000-head California dairy typically spends about $53,600 on energy (2009 dollars), so it would use less than the energy it could generate ($77,500). Consequently, without the ability to sell surplus electricity, farms in California would receive only a fraction of their generated electricity's potential value. In this example, the farm in Wisconsin would have a greater incentive to adopt a biogas recovery system than the farm in California, despite having lower electricity prices.

Table 1. Electricity Use and Price, by Region and Commodity

	Dairy			Hogs		
	Average electricity use per farm (kWh)	Average electricity use per head (kWh)	Average electricity price ($/kWh)	Average electricity use per farm (kWh)	Average electricity use per head (kWh)	Average electricity price ($/kWh)
United States	128,918	1,048	0.069	67,122	158	0.058
West	288,702	893	0.058	7,007	105	0.058
Midwest	101,175	1,102	0.064	64,493	152	0.058
South	159,349	791	0.065	148,651	260	0.055
Northeast	106,418	1,080	0.085	32,264	77	0.072

kWh = kilowatt hour, unit of energy equal to the work done by a power of 1,000 watts operating for 1 hour.

Note: For dairies, a "head" refers to a dairy cow or heifer; for hog operations, a "head" refers to 250 pounds of live weight.

Regions are defined according to U.S. Census of Population and Housing; see www.census.gov/geo/www/us_regdiv.pdf/. Source: USDA, Agricultural Resource Management Survey, NASS and ERS; for hogs, 2004, and for dairy, 2005.

One potential problem with using electricity from biogas facilities is that the onfarm quantity generated may not match onfarm electricity requirements. The quantity generated may fluctuate over the day, month, or year depending on temperature, inflows of manure, machine malfunctions, etc. Similarly, onfarm electricity use fluctuates over time. “Net metering” laws mitigate this problem to a large extent by allowing small-scale generators to obtain the full retail value for the electricity they generate. Under net metering laws, when surplus electricity is produced onfarm, the electricity meter spins backwards, effectively “saving” the electricity until it is needed. Over the billing period, the operation is only billed for its net electricity usage. Recently there has been a rapid adoption of net metering laws, and these laws are currently on the books in more than 40 States (DSIRE, 2010).

Operations that generate more electricity than they consume over a billing period may be able to sell their surplus electricity to the utility at a negotiated price. The price received for this electricity (the selling price) may be different from the price at which they buy electricity (the retail price). The ability to sell surplus electricity and the selling price vary regionally. Recent laws and trends suggest that an increasing number of livestock operations may be able to sell electricity at retail or higher prices. Since manure-derived electricity is from a renewable source, the negotiated price for surplus electricity could enjoy a substantial premium over the wholesale price (the price that utilities pay for electricity from large-scale generators). About 30 States require utilities to purchase a share of their power from renewable sources, including biogas systems (U.S. DOE, 2009).

Figure 1 illustrates the NPV of a methane digester on a California dairy under different assumptions about the electricity price. Data for figure 1 and the following analyses are generated using the model of digester profits that is described in detail in the appendix. The curves in the figure slope upward showing that the returns per head from operating a digester increase with the size of the operation. We discuss the reasons for and implications of the increasing returns to scale in the next section. Operators for whom a digester has a positive NPV could be expected to install a digester, while those for whom the project has a negative NPV would not.

For all the scenarios shown in the figure, it is assumed that operations can obtain the full retail value for their electricity up to the amount of electricity they use onfarm. In other words, we are effectively assuming that operations operate under “net metering” laws. The blue line shows the case where the selling price of electricity equals the retail price. In this case, farms having more than about 1,561 head would earn positive profits from adopting a methane digester. The brown line shows how the NPV changes if electricity prices increase by 15 percent, perhaps because of climate change policy. In this case, the electricity costs saved and the potential revenue from electricity sales would both increase, making digester technology more profitable at all farm sizes. At the higher electricity price, the size at which farms would earn profits from adopting a digester would fall to about 944 head.

Figure 1 also shows that the price of surplus electricity can have a substantial effect on digester profits and adoption. Data from the 2005 ARMS dairy survey indicate that California dairies on average consume 86 percent of the electricity that they could generate from digesters. The additional 14 percent of the electricity generated could be sold off-farm, but if the price were only 50 percent of retail (green line), this reduces the economic feasibility of digesters for smaller operations. In this scenario, no California dairies below about 2,058 head would find it profitable to adopt a digester. Profits would be even lower if dairies had no ability to sell surplus electricity to the grid (effectively setting the selling price to zero). In this scenario (red line), no California dairies below 2,816 head would find existing digester technology profitable.

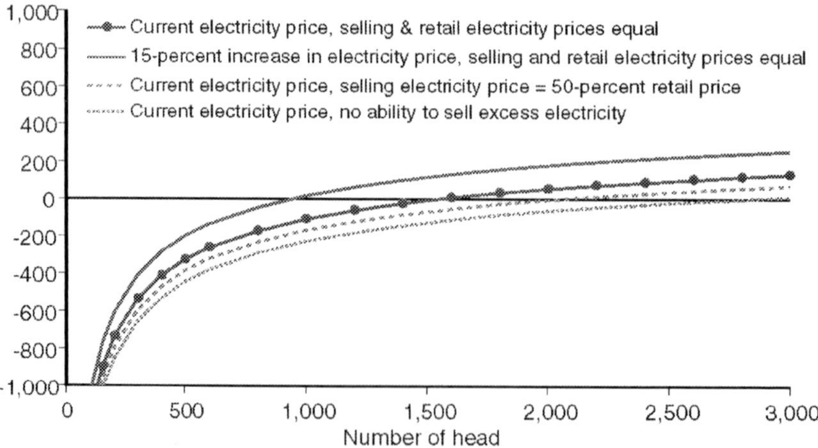

Note: Profits are per head. Legend shows different electricity price and policy scenarios.

Source: USDA, Economic Research Service.

Figure 1. Net present value of profits with a methane digester installed on a California dairy (electricity prices).

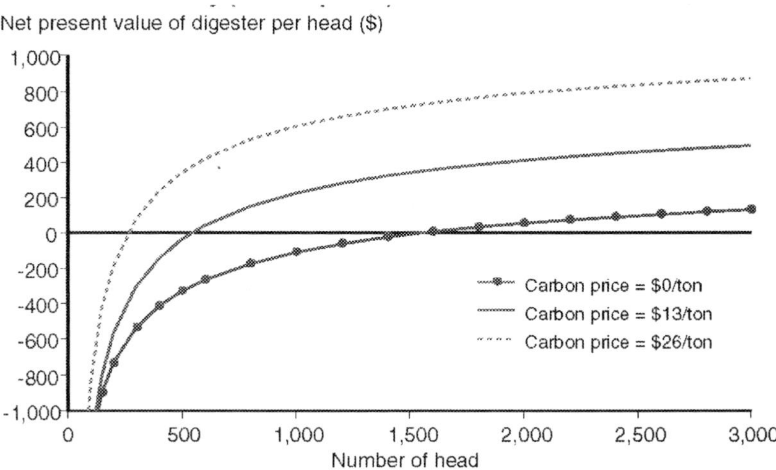

Note: Profits are per head. Legend shows different carbon price scenarios.
Source: USDA, Economic Research Service.

Figure 2. Net present value of profits with a methane digester installed on a California dairy (carbon prices).

Existing Types of Manure Storage and Handling Favor Southern Farms for Offset-Program Qualification

The revenues that a livestock operation could earn in an offset market depend on the type of manure storage and handling facility that an operation has been using. Offset programs usually require that only emission reductions that are additional to current laws, regulations, or practices qualify to be sold as offsets (U.S. EPA, 2008). A key requirement in most offset programs is documentation of baseline emissions and certification that offsets have led to "additional" emissions reductions. Consequently, only operations that had been using an anaerobic manure storage facility before the creation of an offset market would likely qualify for an offset program.[14] This limits the pool of potential offset market participants to swine and dairy operations with manure ponds, lagoons, or slurry pit systems. Operations with slab or shed manure systems or with no manure storage facilities likely would not generate sufficient methane to satisfy the "additionality" requirements for offset certification.[15] We estimate that up to 42 percent of dairies and 54 percent of hog operations have manure management systems that could qualify for an offset program, with Southern States having a higher prevalence of qualifying

Carbon Price Could Play Major Role in Digester Adoption Decision

The sales of carbon offsets provide a potential second source of revenues from methane digesters. Under a cap-and-trade system, the amount of revenue generated from offsets will depend on the market price of carbon and the amount of methane generated, which is a function of farm size, manure management method, and climate, among other things.

Figure 2 illustrates the present value of the additional profits that could be earned by installing a biogas recovery system on a California dairy with a pit manure storage facility at three different carbon prices. In this scenario we assume equal retail and selling electricity prices. The blue line illustrates the NPV when there is no market for offsets—i.e., when the carbon price is zero. In this case, the NPV is negative for all California dairies with pit-based manure management and fewer than 1,561 head, so we would not expect any farms in this size range to adopt the technology unless there were other significant co-benefits or policies to subsidize costs.[10] In contrast, if the carbon price were $13 per metric ton of carbon dioxide equivalent emissions (brown line), operations with an inventory greater than about 544 head would find it profitable to adopt a biogas system.[11] If the carbon price were to double to $26 per metric ton, then the break-even size falls even lower, to less than 265 head.

The figure illustrates the sensitivity of the decision to adopt a biogas recovery system to the carbon price. At the same time, there is a great deal of uncertainty about the future price of carbon. In the major international compliance markets, carbon offset prices have ranged between $15 and $30 per ton of carbon dioxide equivalent emissions in the last decade. In voluntary markets, prices have generally been somewhat lower: ranging between $5 and $15 per ton. In the United States, offset prices have been much lower. The average price for carbon allowances in the RGGI has ranged between $1 and $3 per ton since its inception in 2008.[12] The CCX carbon price has ranged between $1 and $7 per ton since 2004, but has been trading under $1 per ton since 2009.[13] There is a great deal of uncertainty as to the eventual carbon price under a national cap-and-trade system. The EPA forecasted in 2009 that, in the near term, if the House bill (H.R. 2454) were to be enacted it would have resulted in a price of $13 per ton of carbon dioxide equivalent emissions (U.S. EPA, 2009). However, the carbon price could fall short of or exceed this level over the medium or long term.

methane digester. The brown line shows how the NPV changes if electricity prices increase by 15 percent, perhaps because of climate change policy. In this case, the electricity costs saved and the potential revenue from electricity sales would both increase, making digester technology more profitable at all farm sizes. At the higher electricity price, the size at which farms would earn profits from adopting a digester would fall to about 944 head.

Figure 1 also shows that the price of surplus electricity can have a substantial effect on digester profits and adoption. Data from the 2005 ARMS dairy survey indicate that California dairies on average consume 86 percent of the electricity that they could generate from digesters. The additional 14 percent of the electricity generated could be sold off-farm, but if the price were only 50 percent of retail (green line), this reduces the economic feasibility of digesters for smaller operations. In this scenario, no California dairies below about 2,058 head would find it profitable to adopt a digester. Profits would be even lower if dairies had no ability to sell surplus electricity to the grid (effectively setting the selling price to zero). In this scenario (red line), no California dairies below 2,816 head would find existing digester technology profitable.

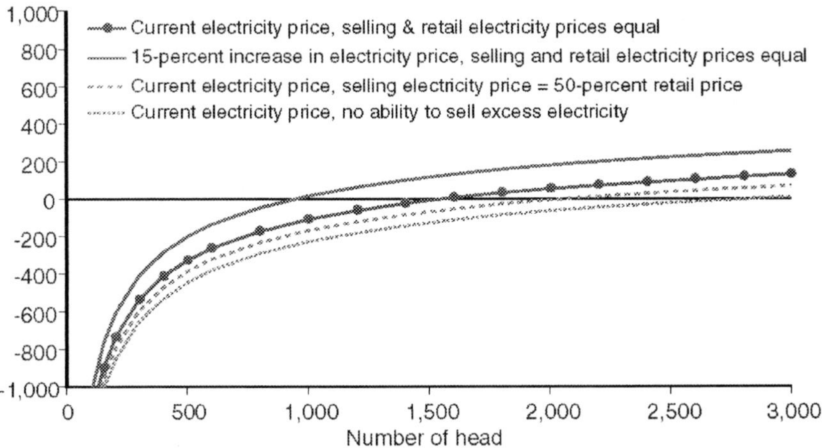

Note: Profits are per head. Legend shows different electricity price and policy scenarios.

Source: USDA, Economic Research Service.

Figure 1. Net present value of profits with a methane digester installed on a California dairy (electricity prices).

One potential problem with using electricity from biogas facilities is that the onfarm quantity generated may not match onfarm electricity requirements. The quantity generated may fluctuate over the day, month, or year depending on temperature, inflows of manure, machine malfunctions, etc. Similarly, onfarm electricity use fluctuates over time. "Net metering" laws mitigate this problem to a large extent by allowing small-scale generators to obtain the full retail value for the electricity they generate. Under net metering laws, when surplus electricity is produced onfarm, the electricity meter spins backwards, effectively "saving" the electricity until it is needed. Over the billing period, the operation is only billed for its net electricity usage. Recently there has been a rapid adoption of net metering laws, and these laws are currently on the books in more than 40 States (DSIRE, 2010).

Operations that generate more electricity than they consume over a billing period may be able to sell their surplus electricity to the utility at a negotiated price. The price received for this electricity (the selling price) may be different from the price at which they buy electricity (the retail price). The ability to sell surplus electricity and the selling price vary regionally. Recent laws and trends suggest that an increasing number of livestock operations may be able to sell electricity at retail or higher prices. Since manure-derived electricity is from a renewable source, the negotiated price for surplus electricity could enjoy a substantial premium over the wholesale price (the price that utilities pay for electricity from large-scale generators). About 30 States require utilities to purchase a share of their power from renewable sources, including biogas systems (U.S. DOE, 2009).

Figure 1 illustrates the NPV of a methane digester on a California dairy under different assumptions about the electricity price. Data for figure 1 and the following analyses are generated using the model of digester profits that is described in detail in the appendix. The curves in the figure slope upward showing that the returns per head from operating a digester increase with the size of the operation. We discuss the reasons for and implications of the increasing returns to scale in the next section. Operators for whom a digester has a positive NPV could be expected to install a digester, while those for whom the project has a negative NPV would not.

For all the scenarios shown in the figure, it is assumed that operations can obtain the full retail value for their electricity up to the amount of electricity they use onfarm. In other words, we are effectively assuming that operations operate under "net metering" laws. The blue line shows the case where the selling price of electricity equals the retail price. In this case, farms having more than about 1,561 head would earn positive profits from adopting a life of the project (assumed to be 15 years), where the cash flow in each year is discounted to its present value.

The model is parameterized using cost and production information derived from several recent case studies of digesters installed on dairy and hog operations. Other livestock sectors were not considered because of a lack of data and the relatively limited contribution of other sectors to total manure methane emissions (U.S. EPA, 2010b). Electricity price data are drawn from the U.S. Department of Energy, and methane emissions are estimated using State-level emission coefficients based on Intergovernmental Panel on Climate Change (IPCC) methods.

For a livestock operation of a given size, type and location, the model provides an estimate of the NPV of the digester. By estimating NPV for every farm in nationally representative samples of dairy (2005) and hog operations (2004) (using USDA's Agricultural Resource Management Survey (ARMS)), the model predicts which farms would adopt a digester at any given carbon offset price. ARMS is conducted by USDA's National Agricultural Statistics Service in conjunction with the Economic Research Service. By summing the reduction in tons of carbon dioxide equivalent emissions for farms that would adopt a digester (i.e., farms on which a digester has a positive NPV), we can estimate the relationship between the price of carbon and the total level of emissions reduced by methane digesters on dairy and hog operations. The ARMS survey weights allow us to extrapolate these estimates to the national level. The model also is used to estimate how farm revenues change with the carbon offset price and to simulate the effect of surplus electricity prices and Government cost-share policies on the potential supply of carbon offsets. More detailed information about the model specification, case studies, data, and parameters is given in the appendix (p. 32).

Electricity—Prices, Onfarm Use, and Sales of Surplus—Is a Key to Digester Profitability

The price of electricity is a key factor determining methane digester profitability because the price determines the cost savings from farm-generated electricity and the revenues that can be earned from the sale of surplus electricity. Electricity prices vary substantially across the country (table 1). For example, because of different retail electricity prices, the electricity generated in 1 year by a 1,000-head dairy with a pit-based digester would be worth approximately $56,300 (retail) in Wisconsin compared with

$77,500 in California. The amounts were calculated using 2009 electricity prices for the industrial sector—$0.0921 per kilowatt hour for California and $ 0.067 per kilowatt hour for Wisconsin (U.S. DOE, 2010). If farms are able to sell surplus electricity to the grid, then the higher electricity prices provide operators with a greater incentive to adopt biogas collectors.

If operations are unable to sell surplus electricity back to the grid, then the benefits from electricity generation are limited to the avoided onfarm energy costs associated with heating or cooling, drying grain, pumping water, lighting, and operating dairy or other machinery. Onfarm energy expenditures per head also vary widely across regions because of differences in climate and production technologies (see table 1). Data from the 2005 ARMS dairy survey indicate that a 1,000-head dairy in Wisconsin typically spends about $125,700 per year on energy (electricity, natural gas, and propane; amount updated to 2009 dollars), which exceeds what it could generate onfarm ($56,300). In contrast, a 1,000-head California dairy typically spends about $53,600 on energy (2009 dollars), so it would use less than the energy it could generate ($77,500). Consequently, without the ability to sell surplus electricity, farms in California would receive only a fraction of their generated electricity's potential value. In this example, the farm in Wisconsin would have a greater incentive to adopt a biogas recovery system than the farm in California, despite having lower electricity prices.

Table 1. Electricity Use and Price, by Region and Commodity

	Dairy			Hogs		
	Average electricity use per farm (kWh)	Average electricity use per head (kWh)	Average electricity price ($/kWh)	Average electricity use per farm (kWh)	Average electricity use per head (kWh)	Average electricity price ($/kWh)
United States	128,918	1,048	0.069	67,122	158	0.058
West	288,702	893	0.058	7,007	105	0.058
Midwest	101,175	1,102	0.064	64,493	152	0.058
South	159,349	791	0.065	148,651	260	0.055
Northeast	106,418	1,080	0.085	32,264	77	0.072

kWh = kilowatt hour, unit of energy equal to the work done by a power of 1,000 watts operating for 1 hour.

Note: For dairies, a "head" refers to a dairy cow or heifer; for hog operations, a "head" refers to 250 pounds of live weight.

Regions are defined according to U.S. Census of Population and Housing; see www.census.gov/geo/www/us_regdiv.pdf/. Source: USDA, Agricultural Resource Management Survey, NASS and ERS; for hogs, 2004, and for dairy, 2005.

farms (table 2). However, a much larger share of production occurs on qualifying farms, especially for hog operations. We estimate that 60 percent of dairy production and 92 percent of swine production occurs on farms with either pit or lagoon systems.

The type of manure storage facility used determines the baseline methane emitted and, consequently, the quantity of offsets that could be generated and the profits that could be earned from a methane digester. Lagoon systems generally emit higher rates of methane per head than pit systems, and operations in warmer climates emit more than those in cooler climates. About 8 percent of dairies and 14 percent of hog operations have only a lagoon, compared with 31 percent and 38 percent, respectively, with only a pit (about 2 percent of operations have both) (see table 2). While pits are more prevalent, larger operations tend to use lagoons more often so that almost a quarter of all dairy cows and more than third of hogs are raised on operations using only lagoons.

Different types of manure storage facilities also have a range of associated digester construction and operating costs. In general, earthen lagoon digester systems are less costly than complete-mix or plug-fl ow pit systems, which are constructed from concrete or steel. However, lagoon digesters can be less productive in Northern States because of low ambient temperatures in the winter months. In cooler climates, pit systems can be heated to promote anaerobic digestion and methane production. However, emissions reductions can only qualify as carbon offsets if they are below the baseline level and the baseline emissions would not include the additional methane generated from heating the digester.

Figure 3 illustrates how the profits from a biogas recovery system can vary substantially across regions and across manure management systems. The figure compares the NPV per head for dairy operations in the two biggest dairy States, Wisconsin and California, when the carbon offset price is $13 (per ton of carbon dioxide equivalent emissions). For both States, operations with lagoons are more profitable than operations with pit systems, mainly because lagoons have much higher initial emissions and consequently higher offset revenues. Lagoons also have lower construction and operating costs. For operations using the same type of manure storage facility, biogas systems are always more profitable in California compared with Wisconsin. This differential is partly explained by California's warmer climate (which increases methane production) and partly by the higher electricity prices in California.

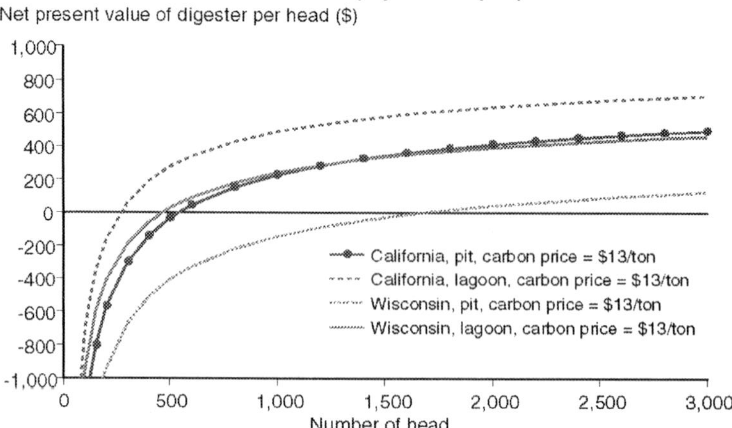

Note: Profits are per head. Legend shows different manure management systems and States.

Source: USDA, Economic Research Service.

Figure 3. Net present value of profits for a methane digester installed on California or Wisconsin dairies (lagoons or pits).

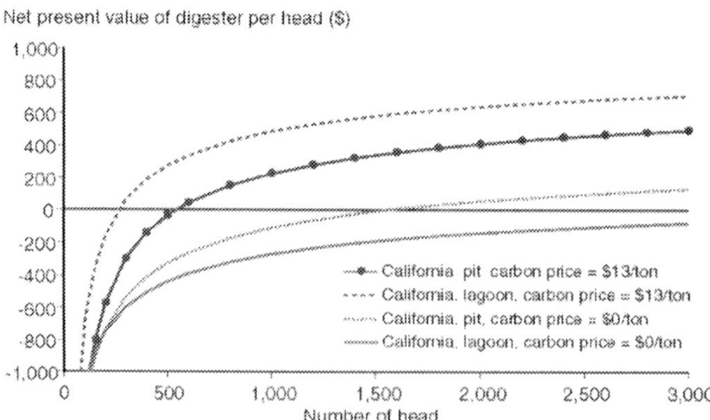

Note: Profits are per head. Legend shows different manure management systems and carbon prices.

Source: USDA, Economic Research Service.

Figure 4. Net present value of profits for a methane digester installed on California dairies (lagoons or pits; carbon prices = $0 or $13 per ton) Net present value of digester per head ($).

Table 2. Percentage of Operations with Different Types of Manure Management Systems, by Region and Commodity

	Total number of farms	Percentage of farms	Percent of farms with...			Percentage of Production	Percent of production with...		
			Lagoon	Pit	Lagoon and pit		Lagoon	Pit	Lagoon and pit
Dairy									
United States	52,237	100.0	8.1	30.9	2.6	100.0	24.7	29.4	5.7
West	6,095	11.7	26.5	18.4	11.6	33.3	36.5	12.2	8.0
Midwest	28,438	54.4	4.3	34.4	1.5	36.4	13.7	45.3	4.8
South	4,034	7.7	22.5	25.9	4.6	9.2	40.5	17.3	8.8
Northeast	13,670	26.2	3.7	30.5	0.2	21.1	18.0	34.4	2.2
Hogs									
United States	40,940	100.0	14.3	37.5	2.3	100.0	35.2	51.9	4.5
West	5,586	13.6	1.0	0.0	1.2	0.6	5.1	0.0	11.8
Midwest	28,539	69.7	9.9	50.3	2.8	71.9	17.7	69.4	5.8
South	5,571	13.6	53.1	4.4	1.3	25.5	87.7	0.9	0.9
Northeast	1,245	3.0	2.1	59.1	1.7	2.0	6.6	84.6	3.1

Note: Regions are defined according to U.S. Census of Population and Housing; see www.census.gov/geo/www/us_regdiv.pdf/. Source: USDA, Agricultural Resource Management Survey, NASS and ERS; for hogs, 2004, and for dairy, 2005.

While pit-based systems are generally more profitable without a carbon market, lagoon-based digester systems become relatively more profitable as the carbon price increases. Figure 4 compares the NPV per head of California pit-versus lagoon-based digester systems with carbon prices of $0 and $13 per ton. Without an offset market (price equal to $0), the pit-based system is more profitable, while at $13 the lagoon-based system has a higher NPV. This is due to the higher baseline methane emissions of lagoons, enabling more revenue from carbon offsets.

Digesters' Other Benefits Include Odor Reduction, Less Surface-Water Contamination

Biogas recovery systems can offer other benefits to livestock producers in addition to electricity generation and marketable emission reductions. Lagoon covers and well-managed anaerobic digestion can substantially reduce odors from manure storage (Welsh et al., 1977; Pain et al., 1990; Wilkie et al., 1995). Digesters can also be designed to reduce the potential for surface-water contamination from pathogens that can be hazardous to animal and human health (Demuynck et al., 1985). By excluding rainwater, a lagoon cover can substantially increase a lagoon's storage capacity and thereby reduce the size or number of lagoons required per operation.[16] An anaerobic digester also can be designed to accept food waste from local food processors or manure from other (local) farms, which can provide additional "fuel" for the digester and a potential source of revenue from "tipping fees" charged to the waste depositors (Bishop and Shumway, 2009).[17] Farms that use a solids separator can use the collected solids onfarm for bedding material or sell them as a soil amendment, which can provide a significant source of income (Leuer et al., 2008).

Scale of Livestock Operations Affects Benefits from Carbon Offsets

There is wide variation in the characteristics of producers who likely would benefit from the introduction of a carbon offset market. The model of digester profits illustrates which producers might benefit from a higher carbon price.

While the costs of installing a biogas recovery system can vary, unit costs generally decline with scale (see the appendix for a discussion of case study data). Scale economies make digester adoption relatively more profitable for larger operations. There are multiple sources of scale economies in biogas recovery and electricity generation. First, the costs of constructing the digester, storage facility, and buildings, which usually comprise the largest component of capital costs, generally decline on a per-unit basis with the size of the operation. Second, the costs of maintaining and repairing the electricity generator and storage facility also tend to decline on a per-unit basis. For example, it usually takes fewer than twice as many hours of labor to monitor or repair a 200-kilowatt (kW) generator compared to a 100-kW generator. Third, there are numerous fixed transactions costs associated with selling electricity or certifying and marketing offsets that do not vary substantially with farm size. Larger operations can spread these fixed costs over a larger revenue base.

Figures 1-4 illustrate the scale economies in manure biogas recovery. They show that profits per head increase for different types of operations. For example, in figure 3, a digester on a 1,500-head Wisconsin dairy with a lagoon manure system (red line) would have an NPV of $333 per head. In contrast, a digester on a 1,000-head dairy in Wisconsin using the same manure management system would have an NPV of only $239 per head. The additional income from biogas systems could enhance existing economies of scale in dairy and hog production.

We next use our model of digester profits to estimate the number, size, and location of operations that would find it profitable to adopt a biogas system at a particular carbon offset price. This provides information about the size and geographic distribution of potential digester adopters under alternative offset market assumptions. In this section we assume that the selling price of surplus electricity equals the retail price.

Figures 5 and 6 illustrate the estimated number and share of U.S. dairy operations in different size categories on which a methane digester would have a positive NPV. Figures 7 and 8 provide the same information for hog operations. Overall, 42 percent of dairies and 64 percent of hog operations have anaerobic manure management systems. The figures show how the number and share of small-scale operations that would potentially adopt a digester increase with the carbon offset price. For all the analyses, the offset price is assumed to be constant over the economic life of the digester.

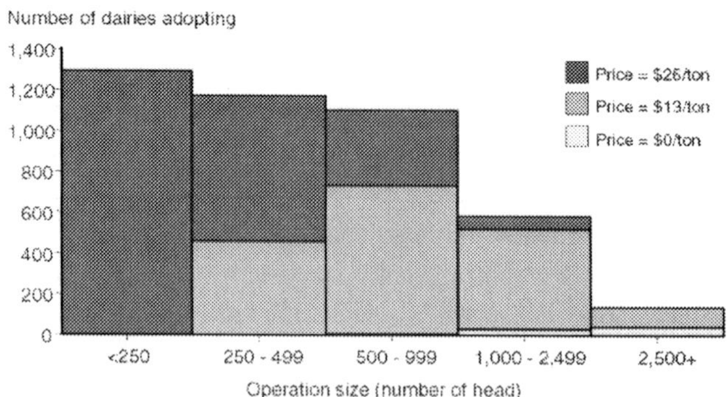

Notes: Numbers at higher prices are additive to those for lower prices; for example, at a price of $1 3/ton, an additional 491 operations of 1,000-2,499 head are predicted to adopt, for a total of 520 operations of this size. At a carbon price of $1 3/ton, no operation smaller than 250 head is predicted to adopt. At a carbon price of $0, no operation with fewer than 500 head and 2 operations of 500-999 head are predicted to adopt.

Source: USDA, Economic Research Service.

Figure 5. U.S. dairy operations on which a methane digester has a positive net present value (three carbon prices).

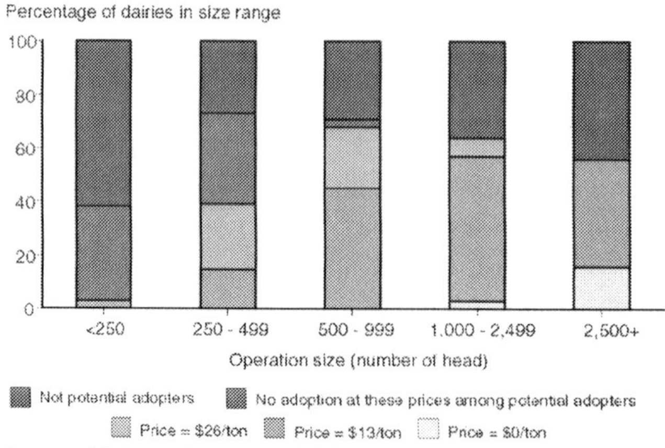

Notes: Numbers at higher prices are additive to those for lower prices; for example, at a price of $13/ton, an additional 54 percent of operations with 1,000-2,499 head are predicted to adopt, for a total of 57 percent of operations of this size. At a carbon price of $13/ton, no operation smaller than 250 head is predicted to adopt. At a carbon price of $0, no operation with fewer than 500 head and 0.1 percent of operations with 500-999 head are predicted to adopt.

Source: USDA, Economic Research Service.

Figure 6. U.S. dairy operations that would profit from methane digesters.

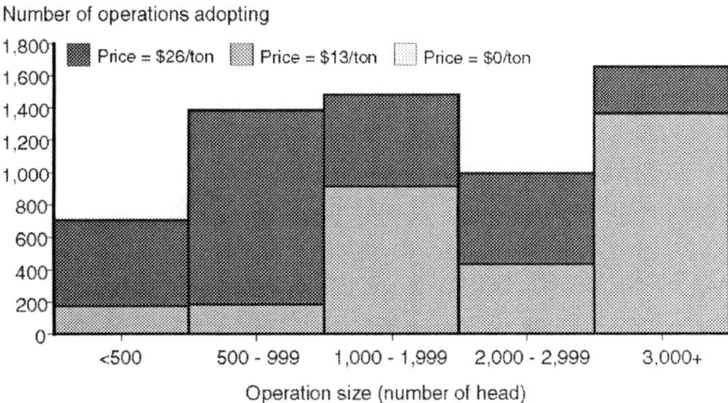

Notes: Numbers at higher prices are additive to those for lower prices; for example, at a price of $26/ton, an additional 557 operations with 2,000-2,999 head are predicted to adopt, for a total of 992 operations of this size. At a carbon price of $0, no operation is predicted to adopt.

Source: USDA, Economic Research Service. Figure 8

Figure 7. U.S. hog operations on which a methane digester has a positive net present value (three carbon prices)

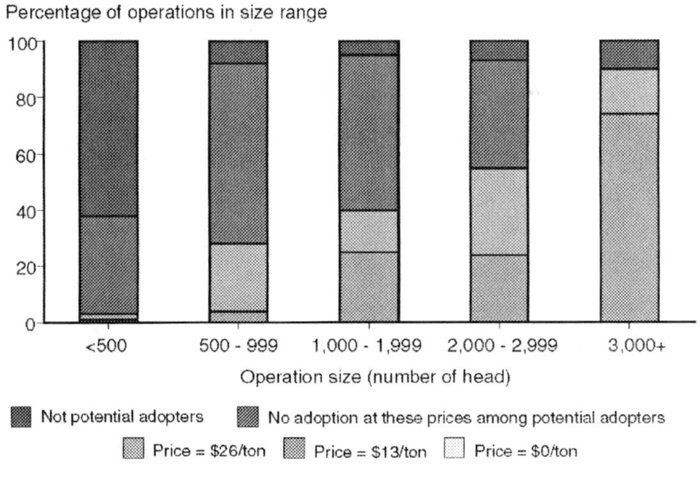

Notes: Numbers at higher prices are additive to those for lower prices; for example, at a price of $26/ton, an additional 31 percent of operations with 2,000-2,999 head are predicted to adopt, for a total of 56 percent of operations of this size. At a carbon price of $0, no operation is predicted to adopt.

Source: USDA, Economic Research Service.

Figure 8. U.S. hog operations that would profit from methane digesters.

With a price of zero (no offset market), digesters on about 40 (or 16 percent of) dairy operations with at least 2,500 head and 29 (or 3 percent of) operations with 1,000-2,499 head would have a positive NPV. Very few smaller operations would have digesters with a positive NPV. However, if the offset price were $13, then digesters on nearly half of operations with more than 500 head and about 15 percent of operations with between 250-499 head would have a positive NPV. If the price were to increase to $26, then the number of digesters with a positive NPV would increase for all farm size categories except the largest (the number of profitable digesters on large farms does not increase because all large farms having either a pit or lagoon manure system already earned profits at the lower price). If carbon offsets could be sold for $26 per ton, then about 1,295 or 3 percent of operations in the smallest size category (< 250 head) would experience positive profits.

For hog operations, the positive relationship between farm size and digester profitability is illustrated in figures 7 and 8. With an offset price of $13 per ton, 74 percent of the operations with more than 3,000 head find a digester profitable compared to only 24 percent of operations with 2,000-2,900 head, 25 percent of those with 1000-1,999 head, 4 percent of those with between 500 and 999, and 2 percent of operations with less than 500 head. The pattern is similar at the higher carbon price of $26.

Without a carbon offset market, no hog operations are estimated to earn positive profits from a digester compared to 71 dairies. This is consistent with the fact that there are currently a much greater number of dairies than hog operations with biogas systems in the United States (U.S. EPA, 2010a). However, at a higher carbon price ($13 per ton), over 7.5 percent of all hog operations would find a digester profitable, compared to 3.5 percent of dairies. This higher rate is partly explained by the fact that a greater share of hog operations have lagoons, which have higher initial methane emissions and are eligible for more offset sales (see table 2).

A substantial share of operations (especially in the smallest size category) lack an anaerobic manure management facility (Figures. 6 and 8). While these operations could construct a digester, it may not be cost effective to do so. Farms that replace an aerobic manure management system (e.g., depositing manure on fields) with a pit or lagoon system would likely not qualify to sell offsets because the resulting emission reductions would not be "additional."

Note: Information for major dairy producing States. Other States indicated by white. ARMS = Agricultural Resource Managment Survery, NASS and ERS.

Source: USDA, Economic Research Service.

Figure 9. Dairy operations that would profit from methane digesters by State, with carbon prices of $13 and $26 per metric ton carbon dioxide equivalent (CO_2e).

Digester Adoption Spreads to More States as Carbon Price Rises

Figures 9 and 10 illustrate the location of dairy and hog operations, respectively, on which digesters are predicted to have a positive NPV when carbon offsets are priced at $13 and $26 per metric ton. The data used to construct the figures are drawn from a 2004 survey of hog producers and a 2005 survey of dairy producers conducted as part of the USDA's Agricultural Resource Management Survey (ARMS). These surveys were conducted in the States accounting for most dairy and hog production.

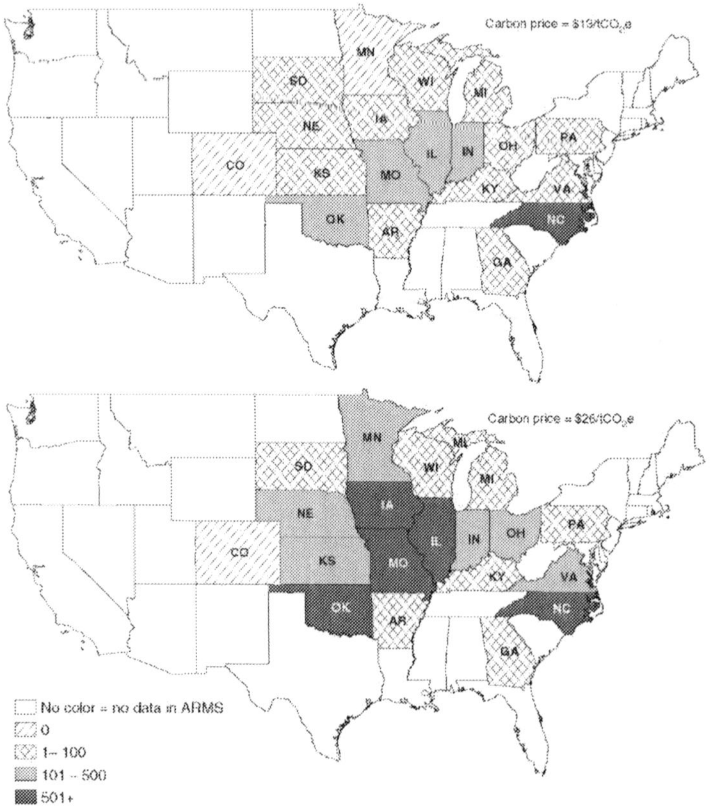

Note: Information for major dairy producing States. Other States indicated by white. ARMS = Agricultural Resource Managment Survery, NASS and ERS.
Source: USDA, Economic Research Service.

Figure 10. Hog operations that would profit from methane digesters by State, with carbon prices of $13 and $26 per metric ton carbon dioxide equivalent (CO_2e).

Figure 9 shows that with an offset price of $13 per ton, there would be over 500 dairies in California, and between 100-500 dairies in New York, Wisconsin, and Texas that would find it profitable to adopt a digester. At a price of $26 per ton, most dairy States see an increase in the number of potential digester adopters. At the higher price, Wisconsin joins California as having over 500 operations that would find it profitable to adopt a digester.

Figure 10 illustrates the geographic distribution of hog operations on which digesters have a positive NPV. At $13 per ton, North Carolina has the largest number of potential adopters. At $26 per ton, Iowa, Illinois,

Missouri, and Oklahoma join North Carolina in having at least 500 potential adopters. There is a somewhat greater share of potential adopters located in Southern States because lagoons are more common in that region. However, the number of potential adopters (shown in Figure 10) does not refl ect this geographic pattern because a large portion of hog production is concentrated in the Midwest.

Options for Promoting Adoption of Biogas Technologies by Smaller Scale Operations

Because of economies of scale, larger producers are likely to benefit most from climate change legislation or other policies that make biogas recovery systems more profitable. The model results suggest that additional profits from biogas systems would tend to enhance existing economies of scale in dairy production (MacDonald et al., 2007) and hog production (Key and McBride, 2008), which could promote further concentration in production over time. In addition, banks may be willing to lend more money to operations with additional digester revenue, which would provide these operations with a greater capacity to expand production over time. However, there are several avenues by which private actions and public sector investments and policies could promote the adoption of biogas systems by smaller scale operations.

Figures 1-6 illustrate how higher profits from biogas recovery (resulting from higher carbon or electricity prices) decrease the size threshold above which it is profitable to adopt the technology. Policies that raise returns will tend to encourage participation by smaller scale operations. Such policies include grants (e.g., USDA Rural Energy for America Program grants), incentive payments (e.g., the U.S. Department of Energy Renewable Energy Production Incentive), tax credits (e.g., the Renewable Electricity Production Tax Credit), accelerated depreciation (Accelerated Cost Recovery System, which allows qualifying renewable energy systems to be rapidly depreciated for tax purposes), property and sales tax exemptions (usually at the State level), and other regulations such as net metering laws and "green" pricing that raise the effective price of electricity. Policies that raise the profitability of biogas recovery systems can be targeted toward smaller scale operations.

Obtaining financing for the large capital investment associated with most biogas systems can be a significant barrier, particularly for smaller scale operations (Gloy and Dressler, 2010). Digesters have little resale value, making their collateral value low. This problem could be addressed by loan guarantee programs such as USDA's Rural Energy for America Program. The uncertainty surrounding digester systems' benefits and costs is another barrier to financing and adoption. Investors who are uncertain about the returns to a project are likely to delay investment or require substantial compensation for the uncertainty (Stokes et al., 2010). Future climate change legislation could increase energy prices and raise carbon offset prices far above current prices in regional carbon trading schemes. However, there is a great deal of uncertainty about the extent and timing of these price increases. Stable and longterm Government policies and programs can help reduce price uncertainty and encourage investment, for example, by providing long-term contracts for carbon offsets and/or electricity.

With high carbon offset prices, it may be possible to design a profitable lower cost biogas system that flares methane rather than using the gas to generate electricity. This approach removes electricity generation from the biogas system, which eliminates the costs of the generator, electrical connections, and maintenance.[18] Such an approach might be economically viable for smaller scale operations that would find it unprofitable to construct and maintain an electricity generator. This option has the greatest potential in the South, where lagoon covers can be installed relatively inexpensively and offer additional benefits to producers such as odor control and rain exclusion.

Centralized digesters used by several livestock producers could allow smaller scale operations to benefit from economies of scale in construction and maintenance. In addition to cost efficiencies, centralized systems could offer benefits in terms of greater marketing leverage in negotiating the sale of electricity, better access to financing, tax credits or grants, and benefits derived from having a manager who could develop specialized skill in digester maintenance and operations (U.S. EPA, 2002).

The main disadvantage of centralized digesters is the additional costs of transporting manure to the centralized facility from the individual farms (Ghafoori and Flynn, 2006). Depending on how the manure is used, there could be additional costs of transporting the digested manure back from the centralized facility to farmland where it can be spread.

Smaller scale livestock operations may be able to achieve a more efficient scale by supplementing manure with food waste products from crop and meat processing facilities, breweries, bakeries, restaurants, etc. (MDA, 2005). When mixed with manure, food waste can provide an efficient feedstock for biogas production and livestock operators can charge "tipping fees" for receiving the waste. However, there is substantial variation in the availability and suitability of food waste for digestion, which may limit the economic and practical feasibility of co-digestion. In addition, there could be additional regulatory requirements associated with handling solid waste, including food waste that could substantially increase costs.[19]

In the hog sector, a large share of finished hog output is produced under production contracts (Key and McBride, 2007). Currently, most production contracts assign growers the responsibility for manure management. However, if digesters become profitable, it is plausible that contractors would seek to share some of the value derived from manure, either directly through a contract or indirectly by reducing the fees they pay farmers to raise hogs. Contractors could potentially facilitate digester adoption on individual operations by helping their contract growers to obtain financing and by providing technical assistance. It is also possible that contractors might establish centralized digesters and require growers to dispose of their manure at a specific facility.

MORE CARBON OFFSETS ARE SUPPLIED AT HIGHER PRICE

As the carbon price rises, so does the potential revenue that can be earned from offset sales. A higher carbon price would likely be associated with an increase in the price of electricity, as most current electricity is generated using "carbon intensive" energy sources. A higher electricity price increases the value of the electricity generated with the methane digester. Hence, a higher carbon price will increase the NPV of digesters and there will be more potential adopters (as was shown in Figures 5 and 7). If we assume that digesters are installed if they have a positive NPV, then we can estimate how an increase in the price of carbon offsets will cause the total greenhouse gas emissions from manure management to drop. By summing the reductions in tons of carbon dioxide equivalent methane emissions from the farms that adopt digesters, we can generate a curve representing the relationship between the price of carbon and the potential quantity of carbon offsets that could be provided.

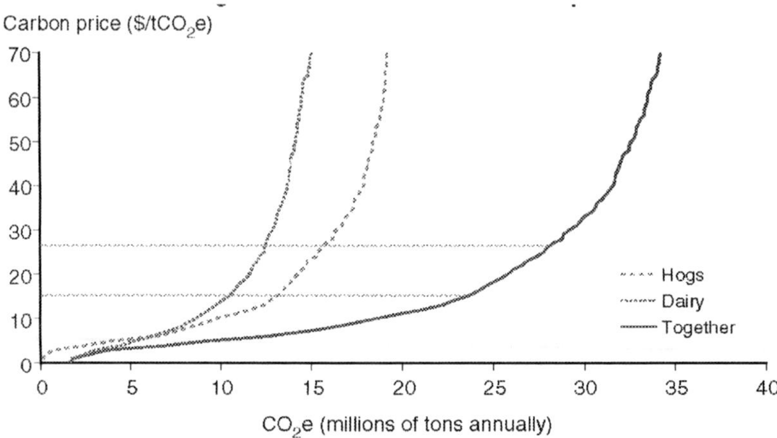

Source: USDA, Economic Research Service.

Figure 11. Total reductions in manure methane emissions from dairy, hogs, and both sectors together at different carbon offset prices.

In our study, we find that the supply of offsets from the dairy sector (brown line, Figure 11) is similar, but somewhat less responsive to the price than was estimated by Gloy (2010). For example, at a price of $15/t$CO_2$e, Gloy estimates dairies would supply 11.6 tCO_2e, compared with our estimate of 10.5 tCO_2e. At a price of $30/t$CO_2$e, Gloy estimates a supply of 16.8 tCO_2e, compared with 12.9 tCO_2e for our study, and at $50/t$CO_2$e, Gloy estimates dairies would supply 19.2 tCO_2e, compared with our estimate of 14.2 tCO_2e. There are several differences between our model and that used by Gloy that could explain these discrepancies.

Figure 11 illustrates this price relationship for dairies, hog operations, and both livestock types combined. The figure shows that without a carbon market (when the price is zero), no hog operations find it profitable to adopt a digester, so there is essentially no reduction in emissions. In contrast, some dairies find that electricity generation alone would make adoption profitable, so some emissions are reduced when the price is zero. As the carbon offset price increases, more and more operations adopt digesters and reduce their emissions. Between about \$2 and \$13 per ton of carbon dioxide equivalent emissions (tCO_2e), the supply of emissions reductions from hog operations increases rapidly. Above a price of about \$13/tCO_2e the supply increases at a slower rate. Eventually, at a price of about \$70 per ton, all the curves approach vertical and the total potential reduction of methane is reached. While dairies supply a larger share of total emission reductions when prices are below about \$6, above that price more of the emission reductions are supplied by hog operations.

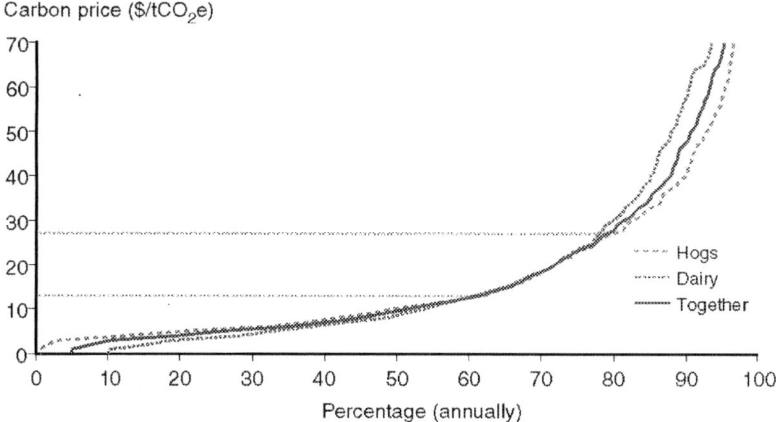

Source: USDA, Economic Research Service.

Figure 12. Percentage reductions in manure methane emissions for dairy, hogs, and both sectors together at different carbon offset prices.

Figure 12 shows the relationship between the carbon offset price and methane reductions as a percentage of the total possible reductions from dairies, hog operations, and both operation types combined. At a carbon price of \$13, greenhouse gas emissions are reduced by 9.8 and 12.4 million tons (carbon dioxide equivalent) for the dairy and swine industries, respectively. This amounts to reductions of 61 percent and 62 percent of manure-generated methane in these industries (or about 62 percent overall). A doubling of the carbon price to \$26 would cause the manure-based methane emissions from dairy and swine together to be reduced by 78 percent.

Higher carbon prices cause the NPV and the present value of electricity cost savings, electricity sales, and offset sales to increase (figure 13). At an offset price of \$13, the dairy and swine industries together could collect a present value of $3.1 billion in offset fees, save $1.2 billion in electricity costs, and earn $361 million in electricity sales (over the 15-year lifespan of the digesters). The present value of total costs are estimated to be nearly $3 billion, yielding an NPV of $1.8 billion at this carbon price over the 15-year lifespan of the digester. As the offset price rises, offset sales contribute an increasingly larger share to digester gross revenues. At a carbon offset price of $5, offset sales contribute only 42 percent of gross revenues, compared with 66 percent when the carbon price is $13, and 76 percent when the carbon price is $26. When the carbon price is low, digester profits are quite sensitive to changes in the price of electricity, as shown in figure 1. However, at higher offset prices, profits are relatively less sensitive to the changes in the price of electricity because electricity sales and savings from avoided costs contribute a relatively small share of profits.

So far we have assumed that operators do not receive any subsidies to help defray the costs of constructing or maintaining the digester. In fact, the operators of virtually all digesters in operation today have received some financial assistance. If farm operators were to receive Government funds for the capital costs of digesters (from a cost-sharing program, for example), digesters would be profitable for more farms, resulting in a greater reduction in emissions and a greater potential supply of carbon offsets. Figure 14 illustrates the percentage reductions in carbon dioxide equivalent methane emissions that could be achieved with different capital cost-share levels at three different carbon prices. Figure 15 shows total Government dollars instead of the cost-share percentage. In the absence of a carbon offset program (a carbon price of $ 0/ton), a 75-percent reduction in manure-based methane from hog and dairy operations could be funded by an 84-percent cost-sharing program in which the Government supplies $4.3 billion. If a carbon offset program existed with a price of $13/ton, then this same reduction could occur with only a 43-percent cost share (or $1.5 billion of Government funds). At a carbon price of $26, no cost sharing would be necessary to induce an emissions reduction of more than 75 percent. Eliminating 95 percent of manure-based methane emissions without a carbon trading scheme could be achieved with a 92-percent cost share and $7.3 billion in Government expenditures. Comparatively, this 95-percent reduction could occur with $6.1 billion (an 84-percent cost share) if the offset price were $13, and $4.8 billion (a 70-percent cost share) if the offset price were $26.

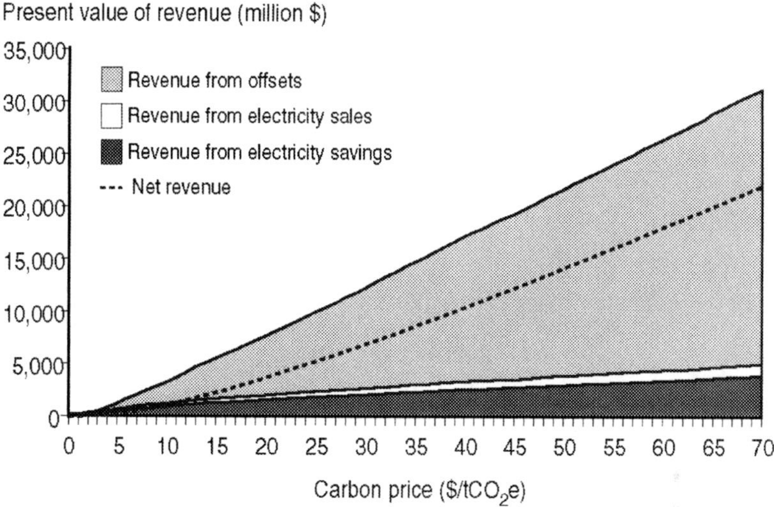

Source: USDA, Economic Research Service.

Figure 13. Gross revenues from methane digester adoption for U.S. dairy and hog industries.

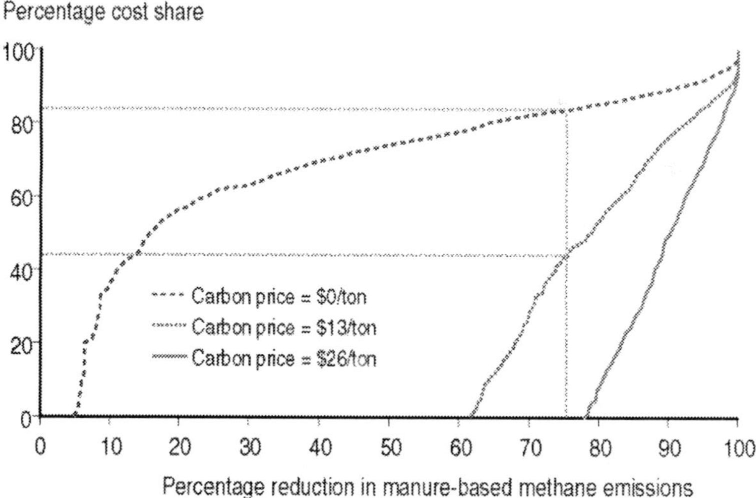

Source: USDA, Economic Research Service.

Figure 14. Percentage reduction in U.S. dairy and hog manure methane emissions by cost-sharing percentage and carbon price.

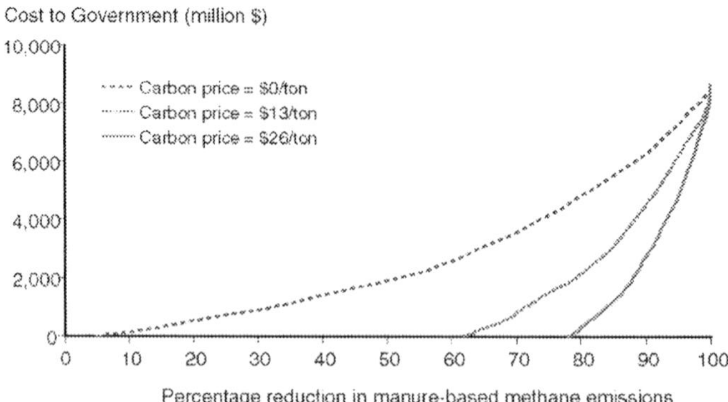

Source: USDA, Economic Research Service.

Figure 15. Percentage reduction in U.S. dairy and hog manure methane emissions by management by cost-sharing program expenditures.

Conclusions

A policy or program that allows producers to be paid for reductions in carbon emissions, such as a carbon offset market or a direct subsidy for emissions reductions, could provide a competitive advantage to U.S. livestock producers who adopt a methane digester. Dairy and hog producers are likely to benefit most from such a policy or program. Among dairy and hog producers, those with high current emissions, particularly those with manure lagoons located in warmer climates, could benefit most from offset sales or emissions reduction subsidies.

Our findings indicate that revenues from the sale of carbon emissions reductions (or offsets) could have a substantial effect on the number of operations that would adopt a methane digester (biogas recovery system). We estimate that a carbon price of $13 per ton would induce dairy and hog operations to supply offsets equivalent to about 22 million tons of carbon dioxide per year. This amounts to about 62 percent of the potential offsets from manure management in these industries, or about 5 percent of total greenhouse gas emissions from U.S. agriculture.

We also found that larger operations would be more likely to participate in a carbon offset market, and likely would earn substantially higher profits per head than smaller scale operations. Among operations with a suitable manure storage facility, only 15 percent of diaries with 250-499 head, compared with 56 percent of operations with more than 2,500 head, could earn positive returns from a digester if the carbon price were $13 per ton. For hog operations, only 4 percent of those with 500-999 head would find it profitable to adopt a digester at that price, compared with 53 percent of operation with more than 3,000 head. Hence, introduction of a carbon offset market could enhance existing economies of scale in production and result in a further concentration of production in the livestock industry among large operations.

There is scope to enhance the adoption of methane digesters on small-scale operations by promoting the provision of long-term contracts for electricity, the construction of centralized digesters (collectively or by a contractor), or the construction of large-scale digesters that take advantage of off-farm food waste. Policies that increase digester profits by increasing the demand for renewable energy, facilitating the ability of livestock operations to sell electricity, or investing in research to lower construction and maintenance costs would encourage greater participation by small-scale operations. Subsidies and tax policies that lower the costs of constructing a digester also could be targeted at smaller operations.

APPENDIX

For this study, we use a model of methane digester profits that depends on farm size and location, electricity and carbon prices, and parameters that characterize the costs and generation capacity of the methane digester. The model is used to estimate digester profits for every farm in two nationally representative surveys of dairy and hog producers.

Recent studies that have modeled the economic benefits and costs of methane digesters include Lazarus and Rudstrom (2007); Leuer et al. (2008); Stokes et al. (2008); and Bishop and Shumway (2009). These studies focus on particular regions, markets, or types of farms, and do not attempt to estimate nationwide digester adoption rates or profits. An exception is Gloy (2010), who develops a general model of digester profitability for dairies in order to estimate the potential supply of carbon offsets from the sector. The modeling approach used in this study is similar to Gloy, but we extend his analysis in several ways by estimating methane digester parameters using case study data, incorporating the hog sector in the analysis, allowing electricity prices to be a function of the carbon offset price, and allowing for transaction costs associated with carbon offset market participation.

We use a discounted cash flow or net present value (NPV) approach to assess the profitability of a digester project. The NPV is the sum of all future cash flows (e.g., revenues from electricity or carbon offsets minus capital and variable costs) discounted to its present value. The NPV approach used in this analysis is deterministic in the sense that real prices are assumed to be known and constant by the operator throughout the economic life of the digester. In fact, many of the benefits and costs associated with a digester are uncertain and variable. For example, both retail prices and selling prices of electricity are likely to fluctuate depending on global economic conditions and policy changes that are difficult to predict. Similarly, carbon offset prices have varied dramatically over time, and estimating future carbon prices is difficult. There is uncertainty about variable costs, which could fluctuate from year to year depending on system reliability and unexpected weather or mechanical failures, as well as uncertainty about methane and electrical output from a digester.

If we knew how much prices and other model parameters are likely to vary, then it would be possible to estimate the probability distribution of the NPV, which would provide a more accurate representation of a digester project's value (Leuer et al., 2008). A further extension could also take into account the irreversible nature of a digester investment. Stokes et al. (2008) use a real option framework to estimate the value to a producer of the option to delay investment in a digester. The authors find that producers would require significant financial compensation, perhaps in the form of assured grant funding or greater electricity prices in order to immediately adopt the technology, rather than delay investment even if the NPV is positive.

By not accounting for the stochastic nature of a digester's benefits and costs nor the option value of delaying investment in this study, we tend to overestimate the value of the digester system and the predicted adoption rates. However, as noted in the text, we also do not account for some possible benefits from a digester such as from "tipping fees" or bedding sales, which causes us to underestimate the value of the project. In addition, we do not account for "nonmarket" benefits from a digester such as odor control or reduced water or air pollution, which also cause us to underestimate the private and social benefits of the project.

The Net Present Value of a Methane Digester Project

An operator who is considering investing in a methane digester has two related decisions: whether to construct a digester that will produce electricity and whether to sell carbon offsets. An operator with a digester will sell offsets if the expected discounted stream of revenues from offset sales exceeds the expected discounted costs of doing so. In other words, the net present value of participating in a carbon offset market NPV_M is positive. The decision to invest in a digester depends on the net present value of electricity NPV_D and of carbon offset sales NPV_M. Hence, there are three possible outcomes:

$$\begin{aligned} &NPV_D \leq 0 \text{ and } NPV_D + NPV_M \leq 0\text{: no investment} \\ &NPV_D > 0 \text{ and } NPV_M \leq 0\text{: construct digester; do not sell offsets} \\ &NPV_D + NPV_M > 0 \text{ and } NPV_M > 0\text{: construct digester; sell offsets} \end{aligned} \tag{1}$$

Another possibility is to construct a digester without an electricity generator and to flare the methane and sell offsets. This scenario is not considered in this study.

The NPV of the digester enterprise for operation i, located in State s, using manure management facility type f is:

$$NPV_D = \sum_{t=0}^{T} \left[(R_{isf} - C_{ift}) / (1+d)^t \right], \tag{2}$$

where T represents the lifespan of the digester, t indexes time, d is the discount rate, R_{isf} is the value of generated electricity (used onfarm and/or sold), and C_{ift} is the cost of constructing and maintaining the digester.

The value of electricity generated by the digester R_{isf} is assumed to be certain and constant over the life of the digester. The value depends on whether the quantity generated onfarm E_i^U

$$R_{isf} = \begin{cases} P_s^{ER} \cdot E_{if}^G & \text{if } E_{if}^G \leq E_i^U \\ P_s^{ER} \cdot E_i^U + P_s^{EW} \cdot \left(E_{if}^G - E_i^U \right) & \text{if } E_{if}^G > E_i^U \end{cases} \tag{3}$$

If the quantity generated is less than or equal to what is used onfarm, then the generated electricity is valued at the buying (retail) price P_s^{ER} If more elec-tricity is generated than is used onfarm, then this surplus electricity $(E_{if}^G - E_i^U)$ is valued at the selling (wholesale) price P_s^{EW} Since the power generation sector is likely to be affected by climate change legislation, we allow the retail and wholesale electricity prices to depend on the carbon intensity of the State energy sources and the price of carbon. Specifically, the retail price of electricity is a function of the observed current retail price P_s^E plus an increase that is proportional to the average carbon dioxide equivalent emissions rate from power plants ϕ_s (in pounds per kilowatt hour (kW/h)) times the carbon price P^M:

$$P_s^{ER} = P_s^E + 0.00045 \cdot \phi_s \cdot P^M, \tag{4}$$

where we multiply by 0.00045 to convert pounds to metric tons.

The selling price of farm-generated electricity will likely also increase with the carbon price. For simplicity, the selling price of electricity is assumed to be proportional to the retail price:

$$P_s^{EW} = \theta \cdot P_s^{ER}, \tag{5}$$

where θ is a parameter that can be varied for policy simulations.

Electricity generation depends on the type of manure storage facility in place and the quantity of manure produced. Generation does not depend on the methane emission factor because emission rates can be increased beyond baseline levels by heating the digester (in pit systems) and by controlling nutrients and liquid/solids ratios. Since the quantity of manure produced is a linear function of the number of head, the quantity of electricity generated can be expressed simply as:

$$E_{if}^G = e_f \cdot N_i. \tag{6}$$

The costs of the biogas system are certain, but vary over the lifespan of the digester. In the initial year ($t = 0$), costs include capital construction costs K_{if} plus maintenance and operating costs V_{if}.20 In subsequent years ($1 < t < T$), costs of the biogas digester only include operating costs:

$$C_{ift} = \begin{cases} \lambda K_{if} + V_{if} \text{ if } t = 0 \\ V_{if} \text{ if } 1 \leq t \leq T \end{cases}. \tag{7}$$

Capital costs include costs of the constructing and designing the pump, pit, heating, building, solids separator, effl uent holder, generator, power lines, and so forth. A share of capital costs ..% is born by a Government cost-share program. The capital costs increase with the scale of the operation at a decreasing rate that depends on parameters a_f and b_f. The cost of this investment is:

$$K_{if} = a_f \cdot (N_i)^{b_f} \tag{8}$$

Annual variable costs V_{if} include costs of maintenance and repairs. Following past studies, we assume that variable costs are proportional to the quantity of electricity generated (which depends on farm size and type of manure handling facility):

$$V_{if} = v \cdot E_{if}^{G} = v \cdot e_f \cdot N_i . \tag{9}$$

The NPV of participating in a carbon offset market is given:

$$NPV_M = \sum_{t=0}^{T} [(P^M \cdot M_{isf} - Z_t) / (1+d)^t], \tag{10}$$

where P^M is the price of carbon offsets (\$/tCO2e), M_{isf} is the quantity of methane that could be sold in the offset market, and Z_t are transaction costs associated with selling carbon offsets.

The quantity of methane produced and burned that would qualify for offset sales is:

$$M_{isf} = N_i \cdot m_{sf} \cdot 24 \cdot 365 \cdot 0.001, \tag{11}$$

where N_i is the number of head and m_{sf} is the methane emission factor (kilogram/methane (kg/CH_4 per head per day)), which is multiplied by 24 (tCO_2e/t CH_4)[21], 365 (days per year), and 0.001 (tons per kg) in order to express in tons of carbon dioxide equivalents (tCO_2e).

Transaction costs associated with selling carbon offsets are certain but vary over the life of the digester. These costs include the initial one-time fixed start-up cost for entering the offset market (Z^E) plus ongoing annual costs of monitoring and verification (Z^V):

$$Z_t = \begin{cases} Z^E + Z^V & \text{if } t = 0 \\ Z^V & \text{if } 1 \le t \le T \end{cases} . \tag{12}$$

Case Studies and Parameter Values

The model parameters, units, and data sources used are shown in table A1. Electricity generation and cost parameters are estimated using information from case studies drawn from three compilations (Lusk, 1998; Kramer, 2004; DPPP, 2006) and from individual articles and reports (Wright and Pershke, 1998; Martin, 2003; Lazarus and Rudstrom, 2007; NNRC, 2007; Michigan Department of Agriculture, 2008; Bishop and Shumway, 2009; Keske, 2009; Moser, undated; Pennsylvania State University, undated). Data from the case studies were used to estimate the parameters if the studies met the following conditions:

1. The case farm was producing heat and/or electricity with its digester technology at the time of the study. This excluded digesters that were constructed solely for odor control or that were nonoperational or had not produced electricity at the time of the case study.
2. The digester was located on an individual farm operation. This excluded digesters at research stations and those that combined manure or other byproducts from multiple sources. This also excluded data collected and synthesized by other researchers or generated by economic models (e.g., Ernst et al., 2009; Crenshaw, 2009; Gloy, 2010; Leuer et al., 2008; Stokes et al., 2008).
3. The case study provided information on the type of digester, specifically whether it was a lagoon- or pit-based system (this excluded Denley and Herndon, 2008).
4. The case study provided startup cost estimates.
5. The name of the farm was provided or the farm could be uniquely distinguished in another fashion (e.g., it was in a State with no other case studies). This was required to avoid double-counting, as several digesters were the subjects of multiple case studies.

We identified 23 case studies of dairies and 11 of hog operations that satisfied our listed conditions. The average farm size, capital and variable costs, and per-head electricity output for the farms used in the analysis are displayed in table A2.

Construction costs per head for the case study operations decline with farm size as illustrated in figure A1. To estimate the cost model parameters (a_f and b_f from equation (8)) we use ordinary least squares and a log-log functional form:

$$\ln(K_{if}) = \alpha_f + \beta_f \ln(N_{if}) + \varepsilon_{if}, \tag{13}$$

where K_{if} is total observed capital construction costs for case study operation i using manure facility type f. The parameters in (8) are computed from the estimated parameters as follows: Separate ragres sions are estimated for the pit and lagoon operations and the estimates are shown in table A3.

For dairies, the coefficient estimates are plausible and statistically significant at the 10-percent level. In contrast, coefficient estimates for the digesters at hog operations are either not statistically significant or are implausible (negative). Given the small sample and insignificant coefficient estimates for the hog operations, we do not use these parameter estimates, but instead use the dairy coefficients adjusted for differences between hogs and dairy cows in terms of the quantity of manure produced per head.[22] Specifically, we use the value from table 1 in Fulhage et al. (2002), which reports that a 1,000-lb dairy cow produces approximately 1.6 times the amount of manure per pound of weight as a 150-lb hog. Hence, we divide the hog inventories that are expressed in pounds by 1,600 to convert into head-of-cow equivalents. For the case studies, we convert number of head of hog to head-of-cow equivalents by multiplying by 150 (to convert to pounds) and then dividing by 1,600.

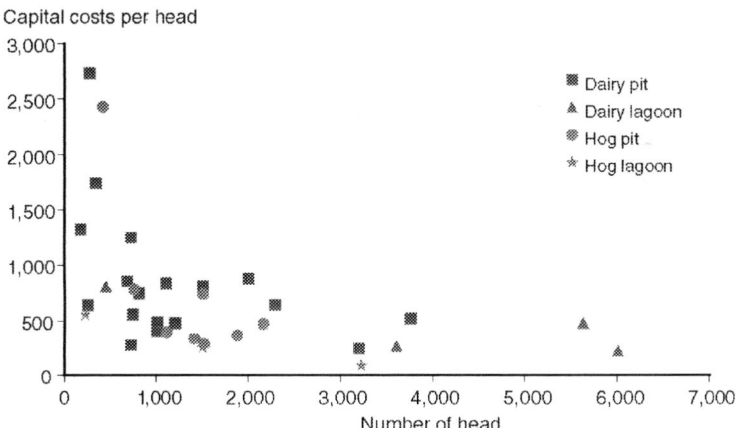

Source: USDA, Economic Research Service.

Figure A1. Capital per head by number of head, dairy and hog operations.

Table A1. Model parameters, values, description, and sources

Variable	Value	Units	Description	Source
Estimated parameters				
$e_{f=pit}$ (dairy)	841	kWh/cow	Electricity produced per dairy cow at an operation utilizing a pit-based digester	Averages based on case studies
$e_{f=lagoon}$ (dairy)	450	kWh/cow	Electricity produced per dairy cow at an operation utilizing a lagoon-based digester	
$e_{f=pit}$ (swine)	41	kWh/hog	Electricity produced per hog at an operation utilizing a pit-based digester	
$e_{f=lagoon}$ (swine)	37	kWh/hog	Electricity produced per hog at an operation utilizing a lagoon-based digester	
v (dairy)	0.033	$/kWh	Variable cost for dairies	
v (swine)	0.014	$/kWh	Variable costs for hog operations	
$a_{f=pit}$	11,708	No unit	Capital investment cost parameter for pit-based digesters	Regression estimates based on case studies
$b_{f=pit}$	0.582	No unit		
$a_{f=lagoon}$	7,864	No unit	Capital investment cost parameter for lagoon-based digesters	
$b_{f=lagoon}$	0.618	No unit		
$P^E{}_S$	Varies by State	$/kWh	State retail electricity price for industrial sector	U.S. DOE, 2010, table 5.6.B

Table A1. (Continued)

Variable	Value	Units	Description	Source
m_{sf}	Varies by State and manure management method	kg CH_4 per head per day	State methane emission factors by manure management method	Chicago Climate Exchange, 2009, tables 3-4
ϕs	Varies by State	lbs/kWh	Carbon emissions factor	US DOE, 2000, table 4
Assumed parameters				
d	0.05	rate	Discount rate	
t	15	years	Economic life of a digester	
T^E	10,000	$	Initial offset market transaction costs	
T^V	3,000	$	Annual offset market transaction costs	
θ	1		Selling price as a fraction of retail price of electricity	
P^M	Varies by policy	$/t CO_2e	Price per ton of CO_2e	
λ	Varies by policy		Percentage of capital costs covered by cost-share program	

kWh = kilowatt hour, unit of energy equal to the work done by a power of 1,000 watts operating for 1 hour.
tCO_2e = per ton of carbon dioxide equivalent.
Source: USDA, Economic Research Service.

Table A58. Averages from Case Studies

Livestock type	Manure management	Cow equivalents	Capital costs ($)	Capital costs per cow equivalent ($)	Variable costs ($/kWh)	Electricity (kWh/year/cow equivalent)
Dairy	Pit	1,195	$750,039	$828	$0.033	841
		(977)	(502,240)	(592)	(0.018)	(382)
Number of case studies		19	19	19	16	15
Dairy	Lagoon	3,916	$1,317,760	$438	$0.033	450
		(2,541)	(977,011)	(264)	(0.018)	(136)
Number of case studies		4	4	4	16	4
Hog	Pit	1,340	$709,255	$719	$0.014	437
		(570)	(281,157)	(713)	(0.005)	(251)
Number of case studies		8	8	8	3	3
Hog	Lagoon	1,648	$264,876	$300	$0.014	391
		(1,503)	(129,366)	(237)	(0.005)	(287)
Number of case studies		3	3	3	3	3

Notes: Standard deviations shown in parentheses. The number of case studies varies according to the number of studies providing information on the specific variable. Variable costs for dairies and hog operations are estimated without dividing between pit and lagoon. All dollar values are in 2009 dollars. Number of head of hog is converted to cow equivalents in the following manner: We assume a 1,000-lb. dairy cow produces approximately 1.6 times the amount of manure per pound of weight as a 150-lb. hog (Fulhage, Pfost, and Schuster (2002), table 1). Hence, we divide the hog inventories that are expressed in pounds by 1,600 to convert into head-of-cow equivalents.

kWh = kilowatt hour, unit of energy equal to the work done by a power of 1,000 watts operating for 1 hour.

Source: USDA, Economic Research Service.

The variable cost parameter from (9) for dairies is a simple average of the 16 case studies for which variable costs are reported (see table A2). Distinguishing the pit versus lagoon systems showed very little difference between the two types, so the same parameter is assigned to both types of systems. Variable costs for hog operations are estimated in the same way using data from four case studies.

Table A3. Regression Estimates: Capital Costs as a Function of Farm Size

Dependent variable: ln(capital)				
	Dairies		Hog operations	
	Pit	Lagoon	Pit	Lagoon
Constant	9.368***	8.970**	13.88***	9.844*
	(0.974)	(1.644)	(2.137)	(1.480)
ln(head)	0.582***	0.618*	-0.0670	0.367
	(0.143)	(0.206)	(0.301)	(0.211)
N	19	4	8	3
R-squared	0.495	0.818	0.008	0.752

Note: Head for hog operations are in cow equivalents, adjusted for weight and manure production. Standard errors shown in parentheses. ***, **, and * refer to significance at the 1, 5, and 10 levels, respectively.

Source: USDA, Economic Research Service.

Table A4. Agricultural Resource Management Survey Hog Categories and Weights

Category	Assigned weight
	Pounds
A sows, gilts, and young gilts bred and to be bred	330
B boars and young males for breeding	330
C cull breeding stock including sows, gilts, and boars	330
D other hogs under 60 pounds	30
E other hogs 60–199 pounds	89.5
F other hogs 120–1 79 pounds	149.5
G other hogs 180 pounds and over	215

Source: USDA, Agricultural Resource Management Survey, National Agricultural Statistics Service and Economic Research Service.

For dairies, the electricity generation parameter e_f is estimated as the average value from the 19 case studies that report this information (see table A2). Pit systems generate substantially more electricity per head than lagoon systems. This is mainly because pit systems are heated in the cooler months to opti-mize methane production, and therefore electricity output. For hogs, we use six case studies to compute the average kilowatt hours per head per year generated by digester systems at pit and lagoon systems. We report these on a per-hog basis in table A1 and on a per-cow-equivalent basis in table A2.

The methane emission factors m_{sf} are based on IPCC tier 2 standards (Chicago Climate Exchange, 2009, tables 3-4). The carbon emissions factors l_s for electricity use by region are from the U.S. Department of Energy's Carbon Dioxide Emissions from the Generation of Electric Power (U.S. DOE, 2000). We assign the same carbon emissions factor to each State within the region.

The model is used to predict digester profits for each farm in the 2005 Dairy and the 2004 Hogs Production Practices and Costs and Returns Reports, which are part of the Agricultural Resource and Management Survey (ARMS). ARMS is a restricted-use dataset conducted by USDA's National Agricultural Statistics Service in conjunction with the Economic Research Service. ARMS contains information on the number and type of animals, type of manure management systems, and costs of electricity consumed.

ARMS Data Description

Farm level data are drawn from the 2005 Dairy and the 2004 Hogs Production Practices and Costs and Returns Reports, part of ARMS. Farms must have sold $1,000 minimum of agricultural products in the prior year in order to qualify for the sample.

Manure Management. ARMS allows farmers to record up to four types of manure storage facilities. For dairies, we classify the following systems as lagoons: "Single stage lagoon (for anaerobic or aerobic digestion)" and "Two stage lagoon (for anaerobic or aerobic digestion in 1st stage, storage in 2nd stage)." We characterize the following as pit-based manure management: "Manure pit (open)," "Manure pit (closed)." The other types of manure storage systems that we do not characterize as either pit or lagoon are "Stacking slab or other open storage of manure," "Slurry or manure tank (open)," and "Slurry or manure tank (covered)."

For hogs, we classify the following systems as lagoons: "Single stage lagoon" and "Two stage lagoon." We characterize the following as pit-based manure management: "Manure pit under building," "Other manure pits (not under building)." The other types of manure storage systems that we do not characterize as either pit or lagoon are "Holding pond (for storage, not anaerobic or aerobic digestion)," "Slurry or manure tank—Open," and "Slurry or manure tank—Closed."

Some farms have both manure and pit systems. In these cases, we discern the percentage of manure held in each type of system, and then use these percentages to weight estimates dependant on the type of manure management (electricity produced, methane produced, and capital costs).

Number of Head. For dairies, ARMS provides the number of head in three categories: Milk cows, dry cows, and breeding bulls. We exclude breeding bulls and find the average number of milk and dry cows over the course of the year.

For hogs, ARMS provides the number of head of various sizes at the beginning and the end of the year. We use these to calculate the total number of pounds of hogs at the operation. We find the average of the beginning and ending number of hogs in each category, and then assign the pounds to each category as shown in table A4. By summing the pounds in each category, we compute the total number of pounds in hogs at the operation. A head is defined as 150 pounds for the computations; however, it is defined as 250 pounds in the figure.

Electricity Use. ARMS records the total amount spent on electricity. We instead need the amount of electricity used. We therefore use the State electricity price from another source (see below) to calculate electricity used in kWh.

State Methane Emissions Factors. The Chicago Climate Exchange provides methane emissions factors by livestock and manure management system for the following categories: (1) Dairy cow, (2) Swine 59 pounds and under, (3) Swine 60 to 199 pounds, (4) Swine 120 to 179 pounds, (5) Swine 180 pounds and above, and (6) Breeding swine. Because ARMS reports the number of hogs in the same swine categories, we use these to calculate specific emissions by hog type. However, there are only two possible methane emissions factors for dairies: "dairy cow" and "dairy heifer." Since ARMS data do not distinguish between dairy cows and dairy heifers (just between milk cows and dry cows), we only use the emission factor for "dairy cows."

REFERENCES

[1] Aillery, M. et al. (2005). *Managing Manure to Improve Air and Water Qual*ity, ERR-9, USDA, Economic Research Service.

[2] Bishop, C. & Shumway, C. R. (2009) "The Economics of Dairy Anaerobic Digestion with Coproduct Marketing." *Review of Agricultural Economics., 31(3)*, 394-410.

[3] Chicago Climate Exchange. (2009). *Chicago Climate Exchange Offset Project Protocol: Agricultural Methane Collection and Combustion*, Accessed at: http://www.chicagoclimatex.com/docs/offsets/CCX_ Agricultural_ Methane_Final.pdf/.

[4] Crenshaw, J. (2009). *What's a digester cost these days?* AgSTAR National Conference, February 24-25, Accessed at: http://www.epa.gov/ agstar/pdf/conf09/crenshaw_digester_cost.pdf/.

[5] Dairy Power Production Program, (2006). *Methane Digester System Program Evaluation Report*, prepared for California Energy Commission Public Interest Energy Research Program. Accessed at: http://communitiesle?p_l_id=59712&folderId=59944&name= DLFE-2429.pdf/.

[6] Database of State Incentives for Renewable Energy (DSIRE). (2010). *Rules, Regulations, and Policies for Renewable Energy*, Accessed at: http:// www.dsireusa.org/summarytables/rrpre.cfm/.

[7] Demuynck, M., Nyns, E. J. & Naveau, H. (1985). "A review of the effects of anaerobic digestion on odour and on disease survival," *Composting of Agricultural and Other Wastes*, J.K.R. Gasser (ed.). Elsevier Applied Science, London, UK, 257-269.

[8] Denley, Jonathan, & Cary, W. Herndon. (2008). *Financial Analysis of Implementing an Anaerobic Digester and Free Stall Barn System on a Mississippi Dairy Farm*, paper presented at the Southern Agricultural Economics Association Meeting, Dallas, TX, February 2-5, 2008.

[9] Ernst, Matthew, et al. (1999). *The Viability of Methane Production by Anaerobic Digestion on Iowa Swine Farms*, Staff Paper No. 328, Iowa State University, Department of Economics.

[10] Environmental Working Group. (2009). *Loopholes in Climate Bill "Offset" Provisions Let Major Polluters off the Hook*, Washington, DC.

Accessed at: http://www.ewg.org/opinion/cap-and-trade

[11] Fulhage, C., Pfost, D. & Schuster, D. (2002). *Fertilizer Nutrients in Livestock and Poultry Manure*, Environmental Quality MU Guide, EQ 351. University of Missouri/Columbia, Outreach and Extension. at: http://extension.missouri.edu/explorepdf/ envqual/ eq0351. pdf/.

[12] Ghafoori, E. & Flynn, P. (2006). *Optimizing the Size of Anaerobic Digesters*, paper presented at the Canadian Society for Bioengineering (SCBE/ SCGAB) Annual Conference, Edmonton, Alberta, Canada. July 16-19, 2006.

[13] Gloy, B. A. (2010). *Carbon Dioxide Offsets from Anaerobic Digestion of Dairy Waste*, working paper WP 2010-05. Cornell University, Department of Applied Economics and Management, Ithaca, NY.

[14] Gloy, B. A. & Dressler, J. B. (2010). "Financial Barriers to the Adoption of Anaerobic Digestion on U.S. Livestock Operations," *Agricultural Finance Review*, 70, 2(2010), 157-168.

[15] Hamilton, K., et al. (2010). *State of the Voluntary Carbon Markets*, report by Ecosystem Marketplace and Bloomberg New Energy Finance, June 14.

[16] Huffstutter, P. (2010). "A stink in Central California over converting cow manure to electricity," *Los Angeles Times*, March 1, 2010. Accessed at: http://articles.latimes.com/2010/mar/01/business/la-fi -cow-power1-2010mar01/.

[17] Intergovernmental Panel on Climate Change. (2007). Chapter 2 "Changes in Atmospheric Constituents and in Radiative Forcing, *Climate Change 2007: The Physical Science Basis*, contribution of Working Group I to the Fourth Assessment Report of the Intergovernmental Panel on Climate Change [S. Solomon et al. (eds.)]. Cambridge University Press, Cambridge, UK. Accessed at: http://www. pcc.ch/pdf/assessmentar4/wg1/ar4-wg1-chapter2.pdf/.

[18] Keske, Catherine, M. (2009). *Economic Feasibility Study of Colorado Anaerobic Digester Projects*, prepared for the Colorado Governor's Energy Office. Accessed at: http://ile.colostate.edu/documents/GEO8-28-9_AD_Economic_Feasibility_by_Keske.pdf/.

[19] Key, N. & McBride, W. (2007). *The Changing Economics of U.S. Hog Production*, ERR-52, USDA, Economic Research Service.

[20] Kramer, J. (2004). *Agricultural Biogas Casebook–2004 Update.* Resource Strategies, Inc., Madison, WI. at: http://www. michigan.gov/ documents/CIS Update_102421_7.pdf/.

[21] Lazarus, W. & Rudstrom, M. (2007). "The Economics of Anaerobic

Digester Operation on a Minnesota Diary Farm," *Review of Agricultural Economics,* 29(2), 349-364.

[22] Leuer, E., Hyde J. & Richard, T. (2008). "Investing in Methane Digesters on Pennsylvania Dairy Farms: Implication of Scale Economies and Environmental Programs," *Agricultural and Resource Economics Review, 37(2)*, 188-203.

[23] Leuer, E., Hyde J. & Richard, T. (2008). "Investing in Methane Digesters on Pennsylvania Dairy Farms: Implication of Scale Economies and Environmental Programs," *Agricultural and Resource Economics Review, 37(2)*, 188-203.

[24] Lusk, P. (1998). *Methane Recover from Animal Manure: The Current Opportunities Casebook.* National Renewable Energy Laboratory, Golden, CO. Prepared by U.S. Department of Energy.

[25] Lusk, P. (1998). *Methane Recovery from Animal Manures: The Current Opportunities Casebook*, ECG-8-17098-01, U.S. Department of Energy, Office of Energy Efficiency & Renewable Energy, National Renewable Energy Laboratory, Golden, CO.

[26] MacDonald, J. M., et al. (2007). *Profits, Costs, and the Changing Structure of Dairy Farming*, ERR-47, USDA, Economic Research Service.

[27] Martin, J. H. (2006). *Comparison of the Performance of Tow Plug-Glow Digesters for Dairy Cattle Manur*e, paper presented at the 2006 AgSTAR National Conference, April 24-26, Madison, WI.

[28] Martin, John, H. (2003). *An Assessment of the Performance of the Colorado Pork, LLC, Anaerobic Digestion and Biogas Utilization System*, paper submitted to Karl Roos, AgSTAR Program, U.S. Environmental Protection Agency, Washington, DC. EPA Contract No. 68-W7-0068 Task Order 5009.

[29] Mehta, A. (2002). *The Economics and Feasibility of Electricity Generation Using Manure Digesters on Small to Mid-Size Dairy Farms*, University of Wisconsin/Madison, Department of Agricultural and Applied Economics, Energy Analysis and Policy Program.

[30] Michigan Department of Agriculture, (2008). *Michigan's Agriculture Industry 2008 Ag-Innovation Grants*. Accessed at: http://www.michigan.gov/documents/mda/aginnfin_268412_7.pdf/.

[31] Minnesota Department of Agriculture. (2005). *Opportunities, Constraints, and Research Needs for Co-digestion of Alternative Waste Streams with Livestock Manure in Minnesota.* Agricultural Resources Management and Development Division, November 23, 2005. Accessed at: http://www. mnproject.org/pdf/CombinedWasteStreamsReport.pdf/.

[32] Moser, Mark A. No year. *A Dozen Successful Swine Waste Digester*, RCM Digesters, Inc. Berkeley, CA. Accessed at: http://www.rcmdigesters.com/images/PDF/ADozenSuccessfulSwineWasteDigesters.pdf/.

[33] Moser, Mark, A., et al. (2007). *Keep the Neighbors Happy—Reducing Odor While Making Biogas*, U.S. Environmental Protection Agency, AgSTAR Program. Accessed at: http://www.epa.gov/agstar/resources proj_sums.html/.

[34] National Research Council. (2003). *Air Emissions from Animal Feeding Operations: Current Knowledge, Future Needs.* National Research Council of the National Academies. Accessed at: http://www.epa.gov/ttn/chief/ap42/ch09/related/nrcanimalfeed_dec2002.pdf/.

[35] Nebraska Natural Resources Committee. (2007). Committee Meeting Transcript LB579 LB581, prepared by the clerk of the legislature, transcriber's office. Accessed at: http://www.legislature.ne.gov/Floor Docs/100/PDF/Transcripts/Natural/2007-02-01.pdf/.

[36] Pain, B. F., et al. (1990). "Odour and ammonia emissions following the spreading of anaerobically digested pig slurry on grassland," *Biological Wastes, 34*, 259-267.

[37] Pennsylvania State Department of Agricultural and Biological Engineering. No year. *Pine Hurst Acres Case Study*. Accessed at: http://www. biogas

[38] Point Carbon. (2010). *8.2Gt CO_2e traded in 2009—up 68% on previous year*, press release. Oslo, Norway. Accessed at:http://www.endseurope.com/ docs/100106a.pdf /.

[39] Scruton, Daniel, L., Stanley, A., Weeks, & Robert, S. Achilles. (2004). *Net Metering's Effect on On-Farm Electrical Generation* (ASAE-044136), paper presented at the 2004 Renewable and Distributed Power Generation: Wind Solar and Other Sources Conference, Ottawa, Canada. August, *3*, 2004.

[40] Stokes, J., Rajagoplan, R. & Stefanou, S. (2008). "Investment in a Methane Digester: An Application of Capital Budgeting and Real Options," *Review of Agricultural Economics, 30(4),* 664-676.

[41] U.S. Department of Agriculture. (2009). *A Preliminary Analysis of the Effects of HR 2454 on U.S. Agriculture*, Office of the Chief Economist. Accessed at: http://www.usda.gov/oce/newsroom/archives/releases/ 2009files/HR2454.pdf/.

[42] U.S. Department of Energy. (2000). *Carbon Dioxide Emissions from the Generation of Electric Power in the United States.* Available at: http:// www.eia.doe.gov/electricity

[43] U.S. Department of Energy. (2009). *States with Renewable Portfolio Standards.* Accessed at: http://apps1.eere.energy.

[44] U.S. Department of Energy, U.S. Energy Information Administration, Office of Coal, Nuclear, and Alternate Fuels. (2010). *Electric Power Monthly, June 2010.* DOE/IEA-0226. Table 5.6.B. Accessed at: http://www.eia.doe.gov/ cneaf/electricity/epm/matrix96_2000.html/.

[45] U.S. Environmental Protection Agency. (2002). *Managing Manure with Biogas Recovery Systems: Improved Performance at Competitive Costs.* AgSTAR Program. EPA-430-F-02-004. Available at: *nepis.epa.gov/Exe/ ZyPURL.cgi?Dockey =2000ZL4E. txt/.*

[46] U.S. Environmental Protection Agency. (2006). Market Opportunities for Biogas Recovery Systems: A Guide to Identifying Candidates for On-Farm and Centralized Systems. AgSTAR Program. EPA-430-8-06-004. Available at: nepis.epa.gov/Exe/ZyPURL.cgi?Dockey=P1008 VEI.txt/.

[47] U.S. Environmental Protection Agency. (2008). Climate Leaders Greenhouse Gas Inventory Protocol Offset Project Methodology for Project Type: Managing Manure with Biogas Recovery Systems, Version 1.3. Available at: www.epa.gov/climateleaders/.../ draft

[48] U.S. Environmental Protection Agency. (2009). EPA Analysis of the American Clean Energy and Security Act of 2009 H.R. 2454 in the 111th Congress. June 23, 2009. Available at: http://www.epa.gov/ climatechange/economics/pdfs/HR2454_Analysis.pdf/.

[49] U.S. Environmental Protection Agency. (2010a). Number of Operating Digesters (April 2010), AgSTAR Program Accomplishments. Accessed at: http://www.epa.gov/agstar/accomplish.html#wi/.

[50] U.S. Environmental Protection Agency. (2010b). Inventory of U.S. Greenhouse Gas Emissions and Sinks: 1990-2008. EPA 430-R-10-006. April 15, 2010.

[51] Welsh, F., et al. (1977). "The effect of anaerobic digestion upon swine manure odors," *Canadian Agricultural Engineering*, *19(2)*, 122–126.

[52] Wilkie, A. C., Riedesel, K. J. & Cubinski, K. R. (1995). "Anaerobic

digestion for odor control," Nuisance Concerns in Animal Manure Management: Odors and Flies, H.H. Van Horn (ed). University of Florida, Florida Cooperative Extension Service, Gainesville, FL, 56-62.

[53] Wright, Peter, E. & Stephen, P. Perschke. (1998). *Anaerobic Digestion and Wetland Treatment Case Study: Comparing Two Manure Odor Control Systems for Dairy Farms*, Paper No. 984105, presented at the American Society of Agricultural and Biological Engineers Annual International Meeting, Orlando, FL, March 8-10.

End Notes

[1] A single ton of released methane has the same global warming potential as 25 tons of carbon dioxide (over 100 years). Burning a ton of methane reduces its warming potential to the equivalent of 1 ton of carbon dioxide—a reduction equivalent to eliminating 24 tons of carbon dioxide. The global warming potential of 25 is based on the latest Intergovernmental Panel on Climate Change (IPCC) Fourth Assessment Report (2007). Some other studies and the U.S. Environmental Protection Agency's U.S. Greenhouse Gas Inventory Report use a global warming potential of 21 based on the earlier IPCC Second Assessment Report (1996). While the older value from the Second Assessment Report has been retained in the U.S. inventory calculations so that results are comparable across years (U.S. EPA, 2010b, pp. 7-8), we use the most recent IPCC value in this analysis.

[2] The remaining gas consists primarily of carbon dioxide, plus small amounts of other gases, including hydrogen sulfide, ammonia, and nitrous oxide.

[3] This is the total for the "Agriculture" sector, as defined by the United Nations Framework Convention on Climate Change (UNFCCC). This total does not include emissions from inputs to agricultural production that are attributed to other sectors, including fertilizer production, transportation, and electricity generation.

[4] Livestock also emit methane from enteric fermentation produced during digestion. In 2008, over three times as much methane was released from enteric fermentation as from manure management (U.S. EPA, 2010, table 2-8).

[5] Manure management in the dairy and hog industries accounts for 0.6 percent of total U.S. greenhouse gas emissions.

[6] Poultry (layers and broilers) operations generally have lower associated methane emissions than swine or dairies because poultry manure is often handled aerobically. However, some poultry operations (particularly layers) store manure in pits or anaerobic lagoons, which can result in substantial methane emissions (NRC, 2003).

[7] Burning digester biogas can create and destroy other greenhouse gases in addition to methane and carbon dioxide. Among these other gases, nitrous oxide (with a global warming potential 298 times that of carbon dioxide) may be the most important. Nitrous oxide is emitted from manure handling facilities and is also emitted when biogas is burned. The net effect of burning biogas on net greenhouse gas emissions from gases other than methane and carbon dioxide is not well established in the scientific literature.

[8] Policies that increase the selling price of electricity would increase incentives to adopt a digester, but could result in unintended consequences for GHG emissions. For example, these policies could provide an incentive to convert from an aerobic to an anaerobic digester or to heat a digester so as to increase methane emissions and therefore electricity output. This could potentially increase GHG emissions as more methane is created and burned.

[9] A carbon content tax would be levied at a rate proportional to the quantity of carbon-equivalent GHGs emitted during the production process.

[10] In fact, in 2010 there were 15 operating digesters on dairies in California, of which 3 are pit-based (U.S. EPA, 2010). Many of these operations likely qualified for cost subsidies or enjoyed other co-benefits that were not included in the investment model.

[11] The price \$13/tCO_2e was used by the EPA in its core scenario analysis of H.R. 2454 (U.S. EPA, 2009).

[12] See Regional Greenhouse Gas Initiative, Market Monitor Reports, http://www.rggi.org/market/market_monitor/.

[13] See Chicago Climate Exchange, CXX Carbon Financial Instrument Contracts Daily Report, http://www.chicagoclimatex.com/market/data/summary.jsf /.

[14] Construction of an anaerobic digester on an operation currently producing little methane from its manure handling practice may not result in a net reduction in greenhouse gas emissions. In fact, construction of an anaerobic digester would likely increase emissions on the operation, though the subsequent burning of the methane to generate electricity could offset this increase.

[15] We assume in the model that the initial level of methane emissions determines the "baseline" level of emissions. Emission reductions below the baseline can be sold as offsets. If the number of head on an operation changed over time then program rules would need to establish how the baseline emission levels would change, if at all. This study does not consider scenarios where the number of head of livestock on farms changes over time.

[16] See "Covers for Long-term Dairy Manure Storages, Part 2: Estimating your Farm's Annual Cost and Benefit." at: http://www.ansci.cornell.edu/pdfs/pdFactSheetSC2PDv.pdf/.

[17] In the case study analyzed by Bishop and Shumway (2009), accepting food waste was found to be profitable for the digester owner, while transportation costs made accepting manure from local farms unprofitable.

[18] Costs associated with electricity generation comprise a substantial share of the total cost of biogas systems. For example, Bishop and Shumway (2009, p. 399) report in their case study of a 1,500-head digester that the cost of the generator was about 31 percent of the total capital costs.

[19] Methane from food waste may not be included in the livestock operation's baseline GHG emissions levels. So it is not clear whether burning methane generated from food waste would qualify for carbon offsets—this would depend on offset market rules and whether the initial GHG emissions from food waste decomposition could be included in the baseline.

[20] In this study, we did not explicitly consider costs associated with obtaining air quality permits or costs associated with installing equipment to comply with air quality standards. Recent news accounts suggest that in some regions or States (such as California) these costs could be substantial (Huffstutter, 2010).

[21] Methane has 25 times the global warming capacity of carbon dioxide over 100 years (IPCC, 2007). Burning methane emits 1 ton of CO_2, so burning 1 ton of methane is equivalent to eliminating 24 tons of carbon dioxide.

[22] We assume that among "pit" or "lagoon" manure storage facilities, dairies and hog operations will have similar construction and operating costs, after correcting for livestock manure output and methane content. However, there may be differences in how manure facilities are constructed on hog and dairy operations, which could cause costs to differ. Unfortunately, there is currently insufficient case study information to accurately quantify these differences.

In: Methane Digesters and Biogas Recovery ISBN: 978-1-61324-594-1
Editor: James M. Hayworth

Chapter 3

United States Anaerobic Digester: Status Report

United States Environmental Protection Agency

Introduction

Livestock producers face a significant challenge in managing manure and process water in a way that controls odors and protects environmental quality. Additionally, livestock manure management practices in the United States are estimated to emit about 2 million tons of methane and account for approximately 8 percent of U.S. methane emissions from anthropogenic activities annually (Figure 1). Biogas recovery systems can provide a multitude of benefits including odor control, improved air and water quality, improved nutrient management flexibility, and the opportunity to reduce greenhouse gas emissions and capture biogas—a useful source of energy.

A biogas recovery system is an anaerobic digester coupled with a device that captures and combusts biogas to produce electricity, heat, or hot water. Biogas is produced when the organic matter in manure decomposes anaerobically (in the absence of oxygen). Biogas typically contains 60 to 70 percent methane, the primary constituent of natural gas, and is a clean burning fuel. Due to facilitation by federal, state, and local programs, the number of operating digester projects has increased from about a dozen in 1990 to over 150 today.

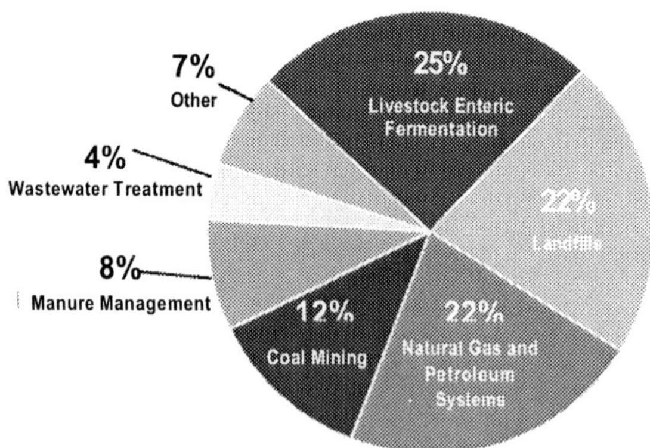

Source: U.S. Environmental Protection Agency, Inventory of U.S. Greenhouse Gas Emissions and Sinks: 1990 – 2008, April 2010.

Figure 1. Human Related Sources of Methane in the United States (percent of total methane emissions).

Table 1. Electricity Generation Potential for Biogas Recovery Systems at Animal Feeding Operations

Animal Sector	Candidate Farms	Energy Generating Potential		
		MW	MWh/year	MMBtu/year
Swine	5,596	804	6,341,527	21,643,632
Dairy	2,645	863	6,802,914	23,218,346
Total	8,241	1,667	13,144,441	44,861,978

More information about biogas recovery potential is provided in AgSTAR's 2010 Market Opportunities for Biogas Recovery Systems report available at www.epa.gov/agstar.

This report is designed to provide a brief overview of the current status of operating digesters at U.S. livestock operations.

Energy Potential

Biogas from manure can be collected and burned to supply on-farm energy needs for electricity or heating. Biogas recovery systems are technically feasible at more than 8,000 U.S. dairy and swine operations. These systems offer a substantial business opportunity to increase farm income by offsetting energy purchases or through the sale of produced energy back onto the electricity grid. Biogas recovery systems at these facilities (Table 1) have the potential to collectively generate more than 13 million megawatt-hours (MWh) per year, which will displace about 1,670 megawatts (MW) of fossil fuel-fired generation on the electrical grid each year. Biogas recovery systems are also feasible at some poultry operations.

USDA Collaboration and Support

Cost is one of the primary obstacles encountered by those wishing to install biogas recovery systems at their facilities. Many successfully operating digester projects have been realized with the assistance of federal grants, most commonly U.S. Department of Agriculture Farm Bill awards.

The Farm Security and Rural Investment Act of 2002 (the 2002 Farm Bill) established the Renewable Energy Systems and Energy Efficiency Improvements Program under Title IX, Section 9006. Under this program, the Secretary of Agriculture provides loans, loan guarantees, and grants to farmers, ranchers, and rural small businesses to purchase renewable energy systems and make energy efficiency improvements. On June 18, 2008, the Food, Conservation, and Energy Act of 2008 (the 2008 Farm Bill) was enacted into law. The 2008 version of the Bill includes Section 9007: Rural Energy for America Program (REAP), which expanded and renamed the program. USDA has pledged to continue financial support of anaerobic digestion systems through REAP grants and loans. Through 2009, USDA has awarded a total of approximately $37.2 million for anaerobic digestion systems.

Secretary Tom Vilsack, USDA and Administrator Lisa P. Jackson, U.S. EPA signing the interagency agreement on May 3, 2010.

On May 3, 2010, USDA and EPA announced a new interagency agreement promoting renewable energy generation and slashing greenhouse gas emissions from livestock operations. The purpose of the agreement is to leverage and utilize each other's expertise and resources to advance deployment of commercially ready anaerobic digestion and biogas use technologies that reduce investment and operational risk to farm owners. With this agreement, USDA's Office of Rural Development and the Natural Resources Conservation Service will work together with EPA to increase the number of biogas systems by finding ways to allow more farmers to tap into this too often neglected resource. The AgSTAR Program has an excellent track record of promoting biogas systems on farms through technical assistance, project assessment, and other collaborative approaches.

PROFILE OF ANAEROBIC DIGESTERS

As of July 2010, EPA estimates that 157 digester projects are operating on commercial scale livestock facilities nationwide (Figure 2). This number, as well as all other data provided in this section, is based on AgSTAR's Anaerobic Digester Database, which comprises information collected from project developers, farm owners, press releases, and a variety of other sources.

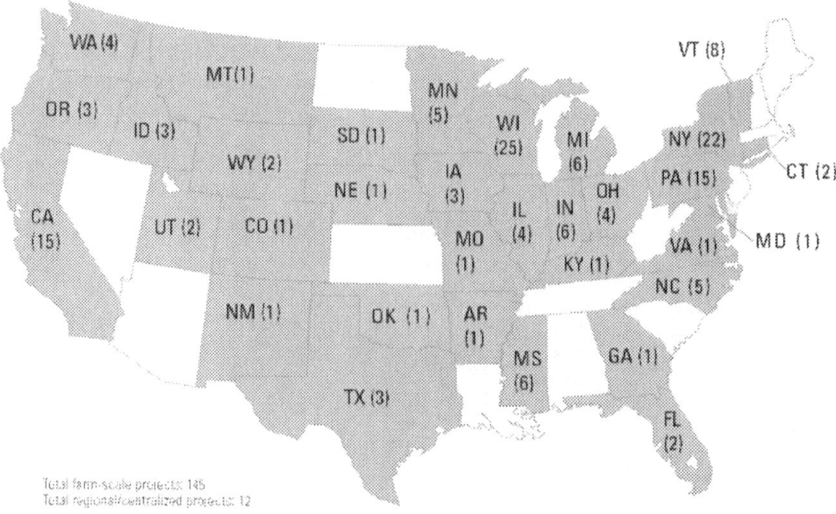

Figure 2. Operating Manure Digester Systems by State (July 2010).

Most digester projects (79 percent) are located at dairy farms, which are largely concentrated in the Midwest, West, and Northeast. Digesters have also been constructed at swine farms (15 percent), beef farms (2 percent), and at poultry farms (3 percent). The poultry farm digester projects are located on 3 broiler farms, 1 layer farm, and 1 duck farm. Table 2 lists the number of projects operating by animal sector.

About half of operating digester projects in the United States use plug flow digesters (Figure 3). Complete mix systems are the second most common digester type, at about 23 percent, followed closely by covered lagoons, at 19 percent. Plug flow digesters are prevalent because this technology is commonly used for scraped manure systems at dairies, and dairy farms currently represent almost 80 percent of the digester projects in the United States.

Table 2. Number of Operating Anaerobic Digester Projects by Animal Type

Farm Type	Total Digester Projects	Plug Flow Projects	Complete Mix Projects	Covered Lagoon Projects	Other Projects
Dairy	126	74	27	16	9
Swine	24	2	5	15	2
Poultry	5	1	4	0	0
Beef	2	2	0	0	0

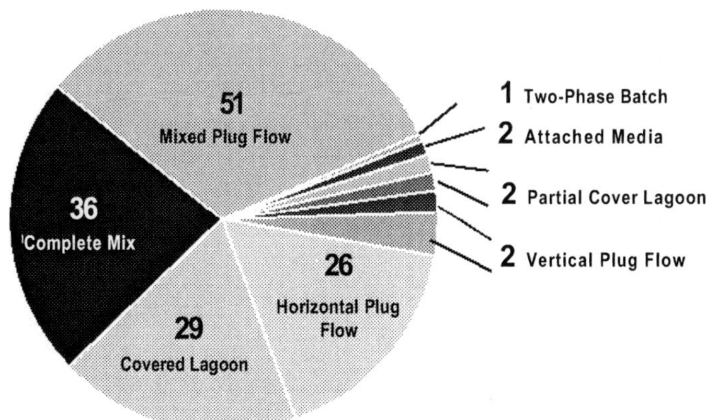

Figure 3. Number of Operating Anaerobic Digester Systems by Technology.

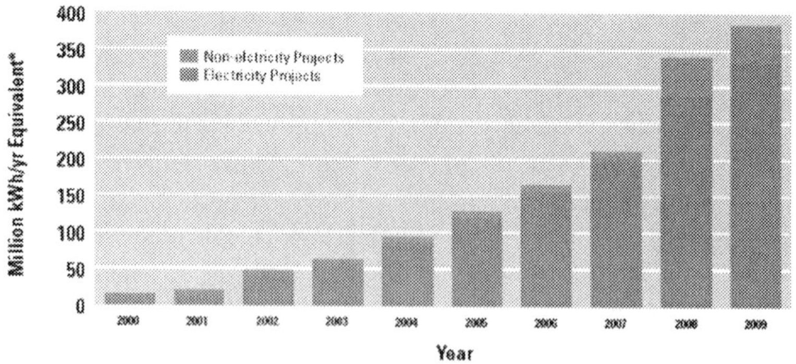

*Output estimated includes a calculated equivalent kWH/yr output for the energy generation of direct thermal, pipeline injection, or other non-electricity-producing projects.

Figure 4. Energy Production by Anaerobic Digester Systems: 2000 through 2009.

Energy Production and Biogas Use

Since 2000, there has been almost a 25-fold increase in the annual electricity generation from digester projects, from about 14 million kilowatt-hours (kWh) to an estimated 331 million kWh per year (Figure 4). Many of the projects that generate electricity also capture engine-generator set waste heat for various on-farm uses in addition to digester heating. Some operations use the biogas as a boiler fuel, process it for injection into natural gas pipeline as renewable natural gas, or flare it for odor control. The percentage of digester projects using the captured biogas for purposes other than generating electricity has increased from nearly zero in 2000 to about 17 percent in 2009. Non-electric energy capture from anaerobic digestion has increased from less than 1 million kWh equivalent in 2000 to about 54 million kWh equivalent of useable energy in 2009.

Profile of Anaerobic Digesters

In 2009, farm digester systems produced an estimated total of 385 million kWh equivalent of useable energy. As of July 2010, total energy production from operating digester projects is estimated to be approximately 404 million kWh/yr, an increase of 19 million kWh over estimates from the end of 2009. This energy production is broken down by state in Figure 5.

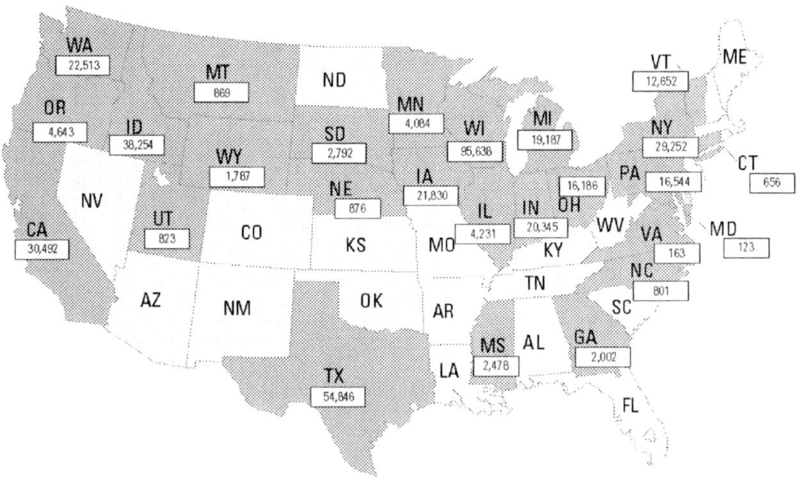

Figure 5. Total Annualized Energy Production of Operating Digesters as of July 2010.

Table 3 shows the breakdown of biogas use among operational digester projects in the United States. Cogeneration projects, which are electricity projects that capture waste heat from electricity generation equipment for on-farm use, are the most common type of digester project. Electricity without cogeneration is the next most popular biogas use, followed by direct use in a boiler or furnace. Less common uses include pipeline injection, vehicle fuel, and methanol production.

Table 3. Biogas Use Technologies for Operational Projects.

Use	Number of Digester Projects*	Percent of Digester Projects (%)
Cogeneration	78	49.7
Electricity	48	30.6
Boiler/Furnace Fuel	20	12.7
Flared Full Time	15	9.6
Unknown Use	7	4.5
Pipeline Gas	4	2.5
Vehicle Fuel	1	0.6
Methanol	1	0.6

*Project totals sum to more than the total number of operating projects because some farms have multiple uses for recovered biogas.

Table 4. Farm Scale, Multiple Farm, and Centralized Projects

Project Type	Number of Operating Projects
Farm Scale	143
Multiple Farm	2
Centralized/Regional	12

Although the majority of systems are farm-owned and operated, using only livestock manure, other approaches are emerging. A dozen centralized systems handling manure from multiple farms are currently operating. Table 4 shows the number of farm scale, multiple farm, and central- ized projects. A multiple farm project refers to a digester system located on one farm that receives influent from a neighboring farm or farms, in addition to influent originating on site. Centralized systems also receive influent from multiple locations but are not located on farms and are typically much larger than farm scale or multiple farm projects. Codigestion, digesting manure with other high strength organic wastes (normally food processing wastes), may significantly increase biogas production per unit volume of reactor. About 22 percent of the currently operating projects codigest other substrates with livestock manure to increase biogas production and revenue.

Emission Reductions

Anaerobic digesters reduce greenhouse gas emissions in two ways. The first is direct methane emission reduction from the capture and burning of biogas that otherwise would escape into the atmosphere from the waste management system. For projects that generate energy, a second benefit is the avoided emissions of greenhouse gases (carbon dioxide, methane, and nitrous oxide) and other pollutants by the use of biogas to displace fossil fuels that otherwise would be used to generate energy. Figure 6 shows the annual emission reductions, including both direct reductions and avoided emissions, resulting from anaerobic digesters since 2001.

In 2009, operating digester projects reduced greenhouse gas emissions by over 1.1 million metric tons of carbon dioxide equivalent. These emission reductions are equivalent[1] to any one of the following annual environmental benefits:

- Removing about 218,000 passenger vehicles from the road.
- Reducing oil consumption by almost 2.7 million barrels.
- Reducing gasoline consumption by over 128 million gallons.

In addition to the benefits of providing an energy source and reduction in greenhouse gas emissions, biogas recovery systems offer a number of air and water quality benefits, including odor control and water quality protection. The primary source of odors from volatile organics and hydrogen sulfide are reduced to methane and carbon dioxide, or contained in the digester and destroyed with the recovered gas. Relative to water quality, anaerobic digestion destroys disease- causing bacteria, and reduces the chemical oxygen demand of the waste that could otherwise result in additional loading on natural waters.

Given the significant benefits recognized from the implementation of anaerobic digestion on agricultural operations and the number of operations that can use these systems across the United States, continued expansion of digester systems is a valuable opportunity for energy and environmental growth.

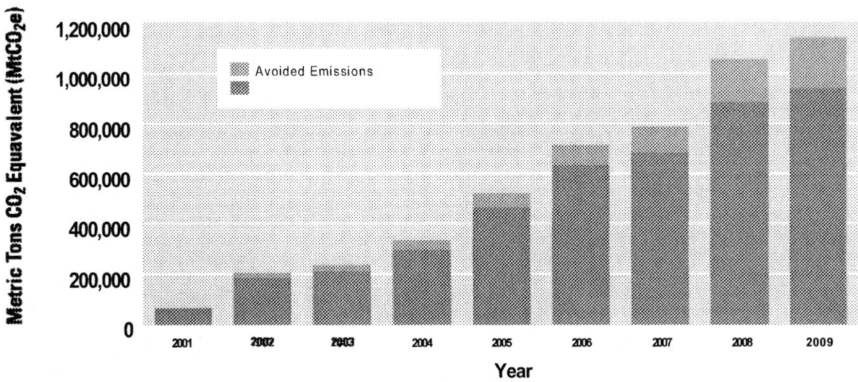

*Avoided emissions calculated based on EPA eGRID national average emission rates for electricity projects and EPA's Inventory of U.S. Greenhouse Gas Emissions and Sinks: 1990-2006 for non-electricity projects. EPA eGRID data was unavailable for 2001, 2002, and 2003, so values were extrapolated based on a linear decrease from 2000 to 2004. EPA eGRID data for 2005 and EPA greenhouse gas inventory data for 2006 were assumed for subsequent years as these are the most recent data available.

Figure 6. Estimated Annual Emission Reductions from Anaerobic Digester Projects*.

End Notes

[1] Emission reduction equivalencies were calculated using EPA's Greenhouse Gas Equivalencies Calculator (March 2010 version) available at www.epa.gov/cleanenergy.

INDEX

D

E

N

O

P

Q

R

S

T

U

V

W